Communication, Culture and Change in Asia

Volume 2

Series editor

Jan Servaes, City University of Hong Kong, Run Run Shaw Creative Media Centre, Hong Kong, China

More information about this series at http://www.springer.com/series/13565

Jan Servaes
Editor

Sustainable Development Goals in the Asian Context

Springer

Editor
Jan Servaes
Bangkok
Thailand

ISSN 2366-4665 ISSN 2366-4673 (electronic)
Communication, Culture and Change in Asia
ISBN 978-981-10-2814-4 ISBN 978-981-10-2815-1 (eBook)
DOI 10.1007/978-981-10-2815-1

Library of Congress Control Number: 2016954013

Printed on acid-free paper

This Springer imprint is published by Springer Nature
The registered company is Springer Nature Singapore Pte Ltd.
The registered company address is: 152 Beach Road, #22-06/08 Gateway East, Singapore 189721, Singapore

Foreword

Communication/Culture and the SDGs: Challenges for a New Generation

2015 will be remembered as a landmark year for global development. Not only for the Paris climate change accord, as momentous as that was, but also for the Sustainable Development Goals, the most ambitious universal development agenda and a significant advance over the Millenium Development Goals. The world has committed, not just to reduce poverty, but to end it, and in so doing to focus, at long last, on the urgent task of addressing inequality so that—to use the key phrase of the new agenda—no one is left behind. And the world has recognized, at long last, the core role of governance—inclusion, participation, transparency, accountability, and human rights—for delivering on that agenda.

This new agenda is particularly important in Asia—where impressive overall growth has come at the expense of much greater inequality, as shown by the fact that although all but four countries in the region have become middle- or higher-income countries, two-thirds of the world's poor still live here.

At UNDP, we find the new agenda strongly aligned with our own vision which is to focus not only on *eradication of poverty* but also *reduction in inequalities and exclusion.* How do culture and communication fit into the new agenda?

The starting point is that the 2030 agenda itself recognizes the important role of cultures for development, noting at paragraph 36 that all cultures and civilizations can contribute to, and are crucial enablers of, sustainable development. One reflection of this is the heightened importance placed on indigenous peoples who are mentioned six times in the agenda.

On the other hand, the world—including our region—is facing a rising challenge of a different sort, which is an attack on human rights and democracy in the name of culture. This is not a new phenomenon of course. Ever since the Universal Declaration of Human Rights itself, there have been attempts from many quarters to undermine the universality and inalienability of human rights under the pretext of respecting so-called cultural or traditional values, and the debate about "Asian values" in our region is just one example of that.

But now this phenomenon is taking an increasingly dangerous form. Not only in Paris, not only in Africa, and the Middle East, but in Myanmar, in Bangladesh, in Indonesia, in Sri Lanka, in the Maldives and beyond, extremist groups are using nationalist and religious language to justify violence and oppression, and to generate conflict between communities. More than ever before, we need to support and show solidarity with those in every country who are proving that the values of the SDGs—peace, solidarity, equality, non-discrimination, inclusion, participation, and tolerance—are truly universal values which underlie every culture and tradition although they are expressed in many different ways. We need to find the translations of these values, humanity's common heritage, across a million different cultural traditions, so that we can overcome the attempt to divide, oppress, and disempower in the name of religion or culture. This then is a global challenge of communication, and it is one in which we in UNDP also engage everywhere we work, and it is one of the ways in which culture and communication are linked.

Communication is central in other ways too. The outcome document explicitly recognizes the great potential of new communication technologies to accelerate human progress and to develop knowledge societies.

But we need to remember that communication is a relationship in which there is not only a speaker but also a listener, and that like all relationships, it is conditioned on power.

We need to ensure not only that the excluded can speak, but also that the powerful are listening. And that conversation will only be meaningful if governments are also communicating by being transparent about policies and laws and their implementation.

That is why we at UNDP particularly welcome the new SDG 16 which covers transparency, accountability, and inclusion, and why we see the SDG as critical to the success of an agenda which is focused on leaving no one behind. But for this goal to be achieved, there is another C that will be critical in addition to communication and culture—and that is civil society. Although civil society is mentioned several times in the agenda, it is not referred to specifically in SDG 16. However, there is no doubt that without a strong civil society, there can be no meaningful participation, inclusion, especially of the most marginalized groups, and there can be no accountability or transparency. For that reason, freedom of expression, association, and assembly are certainly critical among the fundamental freedoms which all member states have undertaken to protect, along with access to information, in SDG target 16.10.

But here again, we are facing a growing challenge. The International Centre for Not-for-Profit Law (ICNL) reports that more than 90 laws constraining freedoms of association or assembly have been proposed or enacted since 2012 around the world, with almost half in Asia and the Pacific. Just as civil society organizations are becoming stronger advocates for the poor and marginalized in this region, the space for them to operate is increasingly shrinking. This is a key priority for us in UNDP, and in August 2015 we organized a round table with civil society from across Asia and the Pacific to consider how to address the problem of civic space in our region.

When we reach the end of this current phase of the journey in 2030, the test of the new agenda will lie in how much the lives of the worst off have been improved, whether we have achieved our aim to eradicate poverty everywhere, whether we have managed to turn around the trend of growing inequality and build greater solidarity across generations and between cultures and communities. The issues that are being discussed in this book, and way that they are reflected in the development policies for this region, will be critical to meeting that test.

UNDP, Bangkok, Thailand Nick Booth

Contents

Editor and Contributors

About the Editor

Jan Servaes (PhD, 1987, Catholic University of Louvain, Belgium) is UNESCO Chair in Communication for Sustainable Social Change. He has taught International Communication and Communication for Social Change in Australia, Belgium, China, Hong Kong, The USA, The Netherlands, and Thailand, in addition to several teaching stints at about 120 universities in 55 countries.

Servaes is Editor-in-Chief of the Elsevier journal "Telematics and Informatics: An Interdisciplinary Journal on the Social Impacts of New Technologies" (http://www.elsevier.com/locate/tele) and Editor of the Lexington Book Series "Communication, Globalization and Cultural Identity" (https://rowman.com/Action/SERIES/LEX/LEXCGC), and the Springer Book Series "Communication, Culture and Change in Asia" (http://www.springer.com/series/13565).

Servaes has undertaken research, development, and advisory work around the world and is the author of journal articles and books on topics such as international and development communication; ICT and media policies; intercultural communication; participation and social change; and human rights and conflict management. He is known for his 'multiplicity paradigm' in "Communication for Development. One World, Multiple Cultures" (1999).

Some of his book titles include: (2016) Servaes, Jan & Oyedemi, Toks (Eds.) *The Praxis of Social Inequality in Media: A Global Perspective.* Lanham, MD: Lexington Books, Rowman and Littefield; (2016) Servaes, Jan & Oyedemi, Toks (Eds.) *Social Inequalities, Media, and Communication: Theory and Roots.* Lanham, MD: Lexington Books, Rowman and Littefield; (2014) Jan Servaes (Ed.). *Technological Determinism and Social Change*. Lanham: Lexington Books; (2014) Jan Servaes and Patchanee Malikhao. Communication for Social Change (in Chinese). Wuhan: Wuhan University Press; (2013) Jan Servaes (Ed.). *Sustainable Development and Green Communication. African and Asian Perspectives*, London/New York: Palgrave/MacMillan; (2013) J. Servaes (Ed.). *Sustainability, Participation and Culture in Communication. Theory and Praxis*. Bristol-Chicago: Intellect-University of Chicago Press; (2008) J. Servaes. *Communication For Development and Social Change*. Los Angeles, London, New Delhi, Singapore: Sage; (2007) J. Servaes & Liu S. (Eds.). *Moving Targets. Mapping the Paths Between Communication, Technology and Social Change in Communities*. Penang: Southbound; (2006) P. Thomas & J. Servaes (Eds.). *Intellectual Property Rights and Communications In Asia*. New Delhi: Sage; (2006) J. Servaes & N. Carpentier (Eds.). *Towards A Sustainable European Information Society*, ECCR Book Series. Bristol: Intellect; (2005) Shi -Xu, Kienpointner M. & J. Servaes (Eds.). *Read The Cultural Other. Forms of Otherness in the Discourses of Hong Kong's Decolonisation*. Berlin: Mouton De Gruyter; (2003) J. Servaes (Ed.). *The European Information Society: A Reality Check*, ECCR Book Series. Bristol: Intellect; (2003) J. Servaes (Ed.). *Approaches To Development. Studies On Communication For Development*, Paris: Unesco Publishing House; and (2002) J. Servaes. *Communication for Development. One World, Multiple Cultures*, Cresskill NJ: Hampton Press. Email: 9cssc9@gmail.com

Contributors

G. Arockiasamy is assistant professor (SG) in visual communication at Karunya University and former head of the department of visual communication at Rathinam College of Arts and Science, Coimbatore, India. His research interests include new media studies, development communication, media culture, and visual communication. He has produced several documentaries. He is the author of book chapters and several articles in peer-reviewed international journals. He has presented several papers in various countries, among them Malaysia and Singapore. His recent project on tribal culture in India is founded by United Board for Christian Higher Education in Asia.

Danesto "Dane" Bacdayan Anacio is currently finishing his PhD in environmental science at the School of Environmental Science and Management (SESAM), University of the Philippines Los Baños in Los Baños, Laguna. His research interest involves the dynamics of cultural, social, and psychological factors with environmental quality and degradation. He is also a member of the UPLB Environmental Science Society (https://www.facebook.com/uplbenvisoc), an organization of graduate students promoting sustainability and environmental science as a practice and philosophy in the Philippines.

Anna Asheervadham Mary is a MSc student of communication in the Department of Visual Communication, Karunya University, Coimbatore, India. She is a very good script writer and an anchor person for television programs.

Nick Booth is the programme advisor on Governance, Conflict Prevention, Access to Justice and Human Rights, at the UNDP Bangkok Regional Hub in Thailand.

Sujeevan Kumar Bullard is a MSc student of communication in the Department of Visual Communication, Karunya University, Coimbatore, India. He is a cinematographer and has produced an ethnographic documentary.

Gaelle Picherit-Duthler (PhD, University of Kentucky) serves currently as the associate dean of the College of Communication and Media Sciences at Zayed University in Abu Dhabi. Prior to this, she served as the graduate program director for the MA in communications (specializations in tourism and cultural communication, and strategic public relations) at Zayed University. She teaches graduate and undergraduate courses within her interests, such as public relations, internal communication, research methods, and corporate social responsibility (CSR). She has published several articles and book chapters on issues of sustainability and CSR, public relations, global virtual teams, and organizational culture. She serves on the board of the GlobCom Institute and is an active member of the Middle East Public Relations Association.

Anthony Le Duc holds a PhD in religious studies. His research is in religious environmentalism. He is presently lecturing at Saengtham College in Thailand and assistant director of Asian Research Center for Religion and Social Communication based in Bangkok. He also serves in various capacities as member of the Society of the Divine Word, a religious congregation in the Catholic Church.

Kamala Vainy Pillai is a lecturer of the public relations major at Curtin University Sarawak Malaysia in Borneo. Her PhD focused on corporate citizenship and minority stakeholder empowerment, specifically indigenous peoples. With over 25 years of experience in various sectors—corporate, civil society, hospitality, and international media, Kamala is an experienced trainer/educator and an ardent advocate of youth participation. She is currently involved in sustainable development-focused community and research projects through ongoing strategic collaboration with corporation, civil society, and community groups.

Pornpun Prajaknate is an assistant professor of communication studies in the School of Communication Arts and Management Innovation at the National Institute of Development Administration (NIDA) in Bangkok, Thailand. Her research interests include development communication, health communication, and information communication technologies (ICT). Her research has focused on development communication, particularly the role of ICT in achieving sustainable development, health promotion, and education. Her current interests mostly revolve around the ICT-based sedentary behaviors of young adults. One of her research articles won the best paper award at the 4th International Conference on Advancement of Development Administration.

Madhushala Senaratne holds a Master's of Arts in Media Practice for Development and Social Change, with distinction, from the University of Sussex, UK, where her thesis examined the changing narrative of development in policy and popular communication, with specific focus on the shifting meanings of inequality and sustainable development. She also holds a Bachelor's of Arts in

International Relations and Media, Communications and Culture, from Keele University, UK, and a postgraduate diploma in conflict and peace studies from the University of Colombo, Sri Lanka. She has previously worked in communications and research in international development and as a journalist in Sri Lanka, and her particular research interests lie in media representations, narratives, and buzzwords in international development, knowledge, power and resistance, and media, conflict and politics in South Asia.

Pavel Slutsky is a full-time associate professor of Communication Management at the Faculty of Communication Arts, Chulalongkorn University, Bangkok. He received his PhD in Political Science (Journalism) in 2008 from St. Petersburg State University, Russia. His research and background drove him toward the areas of political theory of communication (both meta-theory and normative theory), and applied communications (particularly PR and cross-cultural comparative studies). He has published three monographs, three university textbooks, and multiple articles on these topics. He also has over 15 years of experience as an industry practitioner, entrepreneur, and PR-consultant. He is a founder and cofounder of several communication and marketing research agencies which provided services to global corporations such as Walt Disney, Sony, Tuborg, Chanel, Tetra Pak, Audi, and Subaru among others.

Inka Stever is an assistant professor at the College of Communication and Media Sciences at Zayed University in Abu Dhabi and serves as internship coordinator for the college. Her field of expertise is Integrated Strategic Communication. She teaches undergraduate courses such as public relations, writing for integrated strategic communication and capstone projects for seniors. She has published book chapters on issues of sustainability and Corporate Social Responsibility. She comes with over 15 years of work experience in public relations specializing in internal communications and media relations. She received her bachelor of arts degree in politics and history from Ripon College, Wisconsin; her master of arts in North American Studies from the Free University Berlin in Berlin, Germany.

Maya Velmuradova is a PhD graduate and currently an adjunct researcher at the Institute of Research in Sciences of Information and Communication, Faculty of Journalism and Communication Studies of Aix-Marseille University (IRSIC—EJCAM), France. Her research interests include communication of social utility for development and social change (persuasive and participatory), social psychology of communication and of marketing (representations and behavior change, perceived value formation, engagement), and intercultural communication. Her publications have appeared as several book chapters and articles in peer-reviewed journals in French and English. She has been lecturer and assistant professor at South University of Toulon-Var, Aix-Marseille University and University of Perpignan in France, after having been involved in the accompanying of international business and not-for-profit projects in her country of origin, Turkmenistan.

Mallika Vijayakumar is an assistant professor of Communication in the Department of Visual Communication, Karunya University, Coimbatore, India. Her research interests focus on youth, new media, informal learning, and online identity. Her publications have appeared as book chapters as well as journal publications. Dr. Mallika has had a successful stint as an advertising copywriter apart from being associated with academics for two decades.

Katharina Wolf is the coordinator of the public relations major at Curtin University in Western Australia. Katharina has eighteen years of communication and media experience as an educator and practitioner. Her industry experience encompasses communication and research roles in Germany, Spain, the UK, and Australia. Her research interests focus on employability, cross-cultural communication, diversity, and (community) advocacy. Katharina is passionate about providing emerging communicators with a voice and further building the profile of the public relations industry. She is an active member of the Public Relations Institute of Australia, a former state president and chairperson of the institute's national Education Community Committee. Katharina is equally passionate about student-centered, work-integrated learning; a commitment that has been recognised with a number of local and national awards, including a 2011 Australian Learning and Teaching Award for Excellence in Teaching and two Curtin University Awards for Excellence and Innovation in Teaching.

Acronyms

AEC	ASEAN Economic Community
ASEAN	Association of Southeast Asian Nations
BAS	Business Advising Services
BCC	Behavior Change Communication
CCSDG	Communication/Culture and the Sustainable Development Goals
CfD or C4D	Communication for Development
CDSC	Communication for Development and Social Change
CRM	Customer Relationship Manager
CS	Corporate Sustainability
CSO	Civil Society Organization
CSR	Corporate Social Responsibility
CSSC	Communication for Sustainable Social Change
CTCs	Community Technology Centers
DIDIX	Digital Divide Index
EDF	Enterprise Development Project
FAO	Food and Agriculture Organization
FGD	Focus Group Discussions
GAID	Global Alliance for ICT and Development
GATS	General Agreement on Trade in Services
GDP	Gross Domestic Product
GED	Gender Development Index
GEM	Gender Equity Measure
GMMP	Global Media Monitoring Project
GNH	Gross National Happiness
GNP	Gross National Product
GRI	Global Reporting Initiative
GSE	General Self Efficacy
HDI	Human Development Index
HPI	Human Poverty Index
ICNL	International Centre for Not-for-Profit Law

ICT	Information and Communication Technologies
ICTD	Information and Communication Technologies for Development
ILO	International Labor Organization
IMF	International Monetary Fund
ISA	Ideological State Apparatuses
ISCED	International Standard Classification of Education
IT	Information Technology
ITU	International Telecommunications Union
LCR	Learner-to-Computer Ratio
MDGs	Millennium Development Goals
M&E	Monitoring and Evaluation
NGO	Non-Governmental Agency
NICT	New Information and Communication Technologies
NTIA	US National Telecommunications and Information Administration
NWICO	New World Information and Communication Order
OECD	Organization for Economic Cooperation and Development
OPEC	Organization of Petroleum Exporting Countries
PAR	Participatory Action Research
PBS	Public Broadcasting Service
PDR	People's Democratic Republic
PSA	Public Service Advertisement
PVE	Perceived Value of Exchange
PVU	Perceived Value of Use
R&D	Research and Development
RME	Research, Monitoring and Evaluation
RTC	Right to Communicate
SAARC	South Asian Association for Regional Cooperation
SD	Sustainable Development
SDGs	Sustainable Development Goals
SEC	Thai Securities and Exchange Commission
SEU	Subjective Expected Utility value
SME	Small and Medium Enterprise
TAM	Technology Acceptance Model
TBL	Triple Bottom Line
UDHR	Universal Declaration of Human Rights
UK	United Kingdom
UN	United Nations
UNCSTD	United Nations Commission on Science and Technology for Development
UNCTAD	United Nations Conference on Trade and Development
UNDG	UN Development Group
UNDP	United Nations Development Program
UNESCO	United Nations Educational, Scientific and Cultural Organization
UNFCCC	United Nations Framework Convention on Climate Change
UNSC	United Nations Statistical Commission

US	United States
USAID	US Agency for International Development
USSR	Union of Soviet Socialist Republics
WB	World Bank
WCC	World Council of Churches
WCED	World Commission on Environment and Development
WFTO	World Fair Trade Organization
WTO	World Trade Organization

List of Figures

Sustainable Development Goals (SDGs) in Perspective: Lessons from the American Episcopal Missions in Sagada, Northern Philippines

Companies' Accountability in Sustainability: A Comparative Analysis of SDGs in Five Countries

Information Communication Technologies (ICT) for Education Projects in ASEAN: Can We Close the Digital Divide?

A Socio-Cognitive Approach to the Communication of SME Support. An Exploratory Qualitative Study in Turkmenistan

Structural Reform Empowers Sustainable Development and Eradicates Poverty

List of Tables

Introduction: From MDGs to SDGs

Jan Servaes

> *The SDGs represent the change we need to restore people's trust in the global economy. The SDGs represent an enormous opportunity to make our economy work for dignity for all, prosperity for all and a better planet for all. ... The SDGs aim to end poverty in all its forms everywhere, ensuring people live in dignity and prosperity while protecting the planet.*
> (Ban Ki Moon, 17 July 2016).

Abstract This introductory chapter positions the discussion on SDGs within the broader context of development and communication perspectives. After defining 'sustainable development', this chapter first distinguishes between three general development paradigms (modernization, dependency, and multiplicity), and two communication paradigms (diffusion versus participatory communication). Secondly, it identifies the specific communication for social change approaches and strategies which lead towards sustainability. It then evaluates the achievements of the MDGs, and the transition from MDGs to SDGs. While introducing the other contributions, this chapter presents a strong argument for the inclusion of culture and communication in the debate on SDGs.

Keywords Transition from the MDGs to the SDGs · Strengths and weaknesses · History · Realism vs utopia

1 Introduction

The study of communication for development and social change has been through several paradigmatic changes: from the modernization and growth theory to the dependency approach and the multiplicity or participatory model (Servaes 1999). These new traditions of discourse are characterized by a turn toward local communities as targets for research and debate, on the one hand, and the search for an understanding of the complex relationships between globalization and localization, on the other hand.

The early twenty-first-century "global" world, in general as well as in its distinct regional, national, and local entities, is confronted with multifaceted economic and

J. Servaes (✉)
Bangkok 10110, Thailand
e-mail: 9cssc9@gmail.com

J. Servaes (ed.), *Sustainable Development Goals in the Asian Context*,
Communication, Culture and Change in Asia 2,
DOI 10.1007/978-981-10-2815-1_1

financial crises but also social, cultural, ideological, moral, political, ethnic, ecological, and security crises. Previously held traditional modernization and dependency perspectives have become more difficult to support because of the growing interdependency of regions, nations, and communities in our globalized world.

The conclusion we can draw from late-twentieth and early twenty-first-century reconceptualization and reorientations of development and social change is that while income, productivity, and gross national product (GNP) are still essential aspects of human development, they are *not* the sum total of human existence. Just as this has important implications for the way we think about social change and development, so too does it present opportunities for how we think about the role and place of communication and culture in development and social change processes.

We need appropriate strategies to begin addressing these "new" and highly complex challenges. At the 2012 edition of the World Economic Forum in Davos, Switzerland, the background report on the global risks our world faces clearly stated that three common, crosscutting observations emerged from the varied groups of experts consulted:

- Decision makers need to improve understanding of incentives that will improve collaboration in response to global risks;
- Trust, or lack of trust, is perceived to be a crucial factor in how risks may manifest themselves. In particular, this refers to confidence, or lack thereof, in leaders, in the systems which ensure public safety and in the tools of communication that are revolutionizing how we share and digest information;
- Communication and information sharing on risks must be improved by introducing greater transparency about uncertainty and conveying it to the public in a meaningful way.

In other words, more and more one considers communication and culture to be crucial to effectively tackle the major problems of today. Hence, the challenge for us is to revisit and repose the field of Sustainable Development, and to include communication and culture, especially *Communication for Sustainable Social Change* (CSSC), as an essential study field.

Communication for Sustainable Social Change (CSSC) has started to address these specific concerns and issues of food security, rural development and livelihood, natural resource management and environment, poverty reduction, equity and gender, and information and communication technologies (ICTs). However, perspectives on sustainability, participation, and culture in communication changed over time in line with the evolution of development approaches and trends, and the need for effective applications of communication methods and tools to new issues and priorities.

Sustainability and resilience are the two of the many concepts currently "en-vogue" in the academic community, especially with regard to how we understand processes of lasting social change. Indeed, although there is no formal definition of "sustainability," it continues to remain popular in various political, social, and economic discourses, particularly those of environmental groups as a call to action to raise awareness around the current depletion of finite natural resources (for recent overviews, see Agyeman 2013; Grossman 2015; Servaes 2013; Servaes and Malikhao 2014; UN United Nations Secretary-General's High-level Panel on Global Sustainability 2012).

Recently, the term resiliency has been used by a variety of researchers, policy makers, and community organizers as a more relevant supplant to the term sustainability in the context of development and social change. Hopkins (2008: 54) defines resilience as "the capacity of a system to absorb disturbance and reorganize while undergoing change, so as to retain essentially the same function, structure, identity and feedbacks." Zolli and Healy (2012: 7) confirms: "The capacity of a system, enterprise, or a person to maintain its core purpose and integrity in the face of dramatically changed circumstances." Resilience strategies almost always use feedback mechanisms to determine when a disruption is nearing. According to Zolli, a "truly resilient system is able to ensure continuity by dynamically reorganizing both the way it serves its purpose and the scale at which it operates" (p. 10). Thus, strategies of resilience might be developed for economic, social, and ecologic systems. There is no doubt that current global problems such as climate change and the economic crises make the call for "sustainability" and "resiliency" from different groups more urgent. The academic literature, however, has thus far failed to sufficiently explore how groups focused on (environmental) sustainability might themselves become sustainable in the long term.

2 Sustainable Development

Sustainable Development is seen as a means of enhancing decision making so that it provides a more comprehensive assessment of the many multidimensional problems society faces. What is required is an evaluation framework for categorizing programs, projects, policies, and/or decisions as having sustainability potential.

The word is most often associated with being able to meet the needs of the present (socially, economically, environmentally), without compromising the ability of future generations to meet their needs (World Commission on Environment and Development 1987).

Phra Dhammapidhok (Payutto 1998) points out that sustainable development in a Western perspective lacks the human development dimension. He states that the Western ideology emphasizes "competition." Therefore, the concept of "compromising" is used in the WCED definition. Compromising means lessen the needs of all parties. If the other parties do not want to compromise, you have to compromise your own needs and that will lead to frustration. Development will not be sustained if people are not happy.

He consequently reaches the conclusion that the Western perception of and road to sustainability, based on Western ethics, leads development into a *cul-de-sac*.

From a Buddhist perspective, sustainability concerns ecology, economy, and *evolvability*. The concept "evolvability" means the potential of human beings to develop themselves into less selfish persons. The main core of sustainable development is to encourage and convince human beings to live in harmony with their environment, not to control or destroy it. If humans have been socialized correctly, they will express the correct attitude toward nature and the environment and act accordingly. He argues that: "A correct relation system of developed mankind is the acceptance of the fact that human-being is part of the existence of nature and relates to its ecology. Human-being should develop itself to have a higher capacity to help his fellows and other species in the natural domain; to live in a harmonious way and lessen exploitations in order to contribute to a happier world" (Payutto 1998: 189).

This holistic approach of human relates to cultural development in three dimensions:

- Behaviors and lifestyles which do not harm nature;
- Minds in line with (Eastern) ethics, stability of mind, motivation, etc., to see other creatures as companions;
- Wisdom includes knowledge and understanding, attitude, norms, and values in order to live in harmony with nature.

Four dimensions are generally recognized as the "pillars" of sustainable development: economic, environmental, social, and cultural. "The essence of sustainability therefore, is to take the contextual features of economy, society, and environment—the uncertainty, the multiple competing values, and the distrust among various interest groups—as givens and go on to design a process that guides concerned groups to seek out and ask the right questions as a preventative approach to environmentally and socially regrettable undertakings" (Flint 2007: IV).

Over the years, different perspectives—based on both "Western" and "Eastern" philosophical starting points—have resulted in a more holistic and integrated vision of sustainable development. At the same time, a unifying theme is that there is *no universal development model*. Development is an integral, multidimensional, and dialectic process that differs from society to society, community to community, context to context. In other words, each society and community must attempt to delineate its own strategy to sustainable development starting with the resources and "capitals" available (not only physical, financial, and environmental but also human, social, institutional, etc.), and considering needs and views of the people concerned.

Sustainable development implies a participatory, multistakeholder approach to policy making and implementation, mobilizing public and private resources for development and making use of the knowledge, skills, and energy of all social groups concerned with the future of the planet and its people.

Pursuit of this kind of sustainable development requires:

- A political system that secures effective citizen participation in decision making;
- An economic system that provides for solutions for the tensions arising from disharmonious development;

- A production system that respects the obligation to preserve the ecological base for development;
- A technological system that fosters sustainable patterns of trade and finance;
- An administrative system that is flexible and has the capacity for self-correction;
- A communication system that gets this organized and accepted by all parties concerned at all levels of society.

Within this framework, communication and information play a strategic and fundamental role by; (a) contributing to the interplay of different development factors, (b) improving the sharing of knowledge and information, and (c) encouraging the participation of all concerned.

3 One World, Multiple Cultures

The above history has been summarized in "Communication for Development. One World, Multiple Cultures" (Servaes 1999; see also other references in bibliography). We distinguished between three general development paradigms (modernization, dependency, and multiplicity), which were narrowed down to two communication paradigms: diffusion versus participatory communication.

In general, *social change* (or development) can be described as a significant change of structured social action or of the culture in a given society, community, or context. Such a broad definition could be further specified on the basis of a number of "dimensions" of social change: *space* (micro, meso, macro), *time* (short-, medium-, long-term), *speed* (slow, incremental, evolutionary versus fast, fundamental, revolutionary), *direction* (forward or backward), *content* (sociocultural, psychological, sociological, organizational, anthropological, economic, and so forth), and *impact* (peaceful versus violent) (Servaes 2011).

The field of communication for social change is vast, and the models supporting it are as different as their underlying ideologies. In our own work (see Bibliography), we counted *14 different devcom approaches* which currently remain being used and applied. Some of these are more traditional, hierarchical, and linear, some more participatory and interactive. Most contain elements of both. From an epistemological and ontological perspective, that does not always make sense; but in practice that seems to be a given.

Therefore, generally speaking we see two approaches: one aims to produce a common understanding among all the participants in a development initiative by implementing a policy or a development project, that is, the *top-down* model. The other emphasizes engaging the grass roots in making decisions that enhance their own lives, or the *bottom-up* model. Despite the diversity of approaches, there is a consensus in the early twenty-first century on the need for grassroots participation in bringing about change at both social and individual levels.

Development is shaped and done by people—not for people. In order for people to be able to do so, they need to understand "how the system works." Therefore, development or social change should be equated with *empowerment*: the ability of people to influence the wider system and take control of their lives.

It is obvious that people cannot do this entirely on their own. It also requires effort on the part of development change partners (agencies and agents) to help solve some of the dysfunctions in the system and create the enabling conditions. Therefore, this perspective argues that communication needs to be explicitly built into development plans and social change projects to ensure that a mutual sharing/learning process is facilitated. Such communicative sharing is deemed the best guarantee for creating successful transformations.

This new starting point is examining the processes of "bottom-up" change, focusing on self-development of local communities. The basic assumption is that there are no countries or communities that function completely autonomously nor completely self-sufficient and nor are there any nations whose development is exclusively determined by external factors. Every society is dependent on the other in one way or another, both in form and in degree.

4 The MDGs

Launched in 2000 with 2015 as its final deadline the Millennium Development Goals (MDG) initiative follows decades of debate over how nations might collaborate on long-term strategies for a global social agenda. Wealthy countries were asked to increase development aid, relieve the debt burden on poor countries, and give them fair access to markets and technology. While people like Alston (2005) argue that, for development communication, the MDGs "are the most prominent initiative on the global development agenda"; Waisbord (2006: 3) "cannot help but notice that communication goals are absent… While everyone seems to think that communication is important, apparently it is not crucial enough to make it into the (MDG) list." The latter position is confirmed by policy makers within the UN agency system. On June 25, 2007, the director of UNESCO's Bureau of Strategic Planning urged the director of the UN Development Group Office (UNDG) in a letter that

> while Communication for Development remains critical to achieving the MDGs, it has not been adequately recognized as an essential element in development planning within UN programming exercises in general and CCA/UNDAF in particular. Participants expressed the need to rectify this deficiency and to make it a policy to integrate Communication for Development at field level throughout the planning, implementation and evaluation phases of UN assistance programmes.

Each Millennium Development Goal (MDG) has its own set of targets and benchmarks that provide a measurable way to track its implementation (UNDP 2006) (see Annex 1). However, questions can be raised about the feasibility and appropriateness of setting the same global targets for governments worldwide:

> The MDGs can justly claim to generate a bit of buzz about duties a government might otherwise neglect... Sadly, however, they cannot do what they purport to do, which is to provide credible benchmarks against which governments can be judged... Some goals cannot be met, others cannot be measured... The goals are supposed to be everyone's responsibility, which means they are no one's. Poor countries can blame rich ones for not stumping up enough cash; rich governments can accuse poor ones of failing to deserve more money (The Economist 2007: 13).

In essence, development is about the development of people and the transformation of society. The MDGs in general look at development as an "engineering problem" to be solved from a top-down perspective. "Sixty years of countless reform schemes to aid agencies and dozens of different plans, and $2.3 trillion later, the aid industry is still failing to reach the beautiful goal (of making poverty history, JS). The evidence points to an unpopular conclusion: Big Plans will always fail to reach the beautiful goal" (Easterly 2006: 11).

Easterly's critical assessment is not shared by everybody in the MDG debate. One can find four schools of thought in the literature: "Optimists (or perhaps one should say 'architects') see the goals as a vehicle for transforming the human condition (including Jeffrey Sachs, leader of the Millennium Project, Vandemoortele and Pronk); strategic realists see the MDGs as essential to achieving and preserving political commitment (Fukuda-Parr and Jolly); sceptics find the MDGs well-intentioned but badly thought out (such as Clemens and Easterly); radical critics see the MDGs as a diversionary maneuver to draw attention away from the 'real' issues of growing global inequality and gender disparity (including Antrobus, Eyben, Saith, Pogge and Ziai). UNCTAD describes the fundamental problem of the MDGs as 'the lack of a more inclusive strategy of economic development'" (Van Norren 2012: 826). Of course, this division is rudimentary; the position of "optimists" may be explained by their executive responsibility for the design and/or implementation of MDG policy and not per se by a fundamental disagreement on the analysis of development processes.

The position of some has evolved over time. For instance, the comments by Jan Vandemoortele, one of the cowriters of the MDGs who has worked for both UNICEF and UNDP, are interesting in this context:

> The basic criticism levelled against the Millennium Development Goals (MDGs) is that they present a reductionist view of development. They are too limited in scope; their definition is too narrowly focused on the social sectors; their sectoral fragmentation leads to vertical silos; their emphasis on quantification is excessive; and they omit fundamental objectives contained in the Millennium Declaration, such as peace and security, human rights, democracy and good governance. The criticism also points out that they fail to underscore universal values such as freedom, tolerance and equality. Too many dimensions are missing (e.g., human rights and economic growth) or the complexity of the dimensions that are included is missed (e.g., gender equality and quality of education) (Vandemoortele 2011a: 8).

In Vandemoortele (2011b) he presents an insider's view of the origin of the Millennium Development Goals (MDGs). The aim was twofold: to rescue the Millennium Declaration from oblivion and to expand the development narrative beyond economic growth.

> The former has been successful, but not the latter. Since its establishment, the MDG agenda has been permeated with the idolatry of literalism and sanitized to fit the conventional development paradigm. Statistics have been abused to fabricate evidence of success. The great paradox is that poverty is increasingly regarded as a multi-dimensional phenomenon whilst its quantification remains essentially one-dimensional, which reinforces a money-metric perspective of the MDGs. The agenda has been cut back to a standard set of macroeconomic, sectoral or institutional reforms of a technical nature. However, the MDG agenda implies fundamental transformations in society, which are invariably driven by domestic politics and local actors. The world is off track, not because of insufficient economic growth but mostly because people in the bottom quintiles have benefited disproportionately little from national progress. As long as the world continues to turn a blind eye to the growing inequities within countries, the MDGs will be mission impossible (Vandemoortele 2011b: 1).

In the UN system conflicts seldom concern the MDGs as such, but rather the means of achieving them. Most scholars and policy makers agree that the MDGs require fundamental transformations in society.

Therefore, the MDGs should be assessed from a *people's perspective*. In other words: What can MDGs do for the poor and "voiceless"? "Setting a prefixed (and grandiose) goal is irrational because there is no reason to assume that the goal is attainable at a reasonable cost with the available means" (Easterly, ibid.). It is therefore essential to start from the perspective of local communities and to cooperate with organizations (UN, governmental, NGOs, the public and the private sector, and civil society) that have developed a trust within a community in order to achieve sustainable change in society. Sen (2004: 20) aptly summarizes the discussion: "The deciding issue, ultimately, has to be one of democracy. An overarching value must be the need for participatory decision making on the kind of society people want to live in, based on open discussion, with adequate opportunity for the expression of minority positions."

5 Evaluating the MDGs

The MDGs were evaluated on a regular basis. Countries were encouraged to update and improve their performance. The reports suggest that a number of developing countries have made considerable progress toward the accomplishment of certain MDGs, but the overall progress was sporadic across goals, countries, and regions.

Some of the "successes" are as follows:

- The world has reduced extreme poverty by half;
- Efforts in the fight against malaria and tuberculosis have shown results;
- Access to an improved drinking water source became a reality for 2.3 billion people;
- Disparities in primary school enrollment between boys and girls are being eliminated in all developing regions;
- The political participation of women has continued to increase;

- Development assistance rebounded, the trading system stayed favorable for developing countries, and their debt burden remained low;
- Major trends that threaten environmental sustainability continue, but examples of successful global action exist;
- Hunger continues to decline, but immediate additional efforts are needed to reach the MDG target;
- Chronic undernutrition among young children declined, but one in four children is still affected;
- Child mortality has been almost halved, but more progress is needed;
- Much more needs to be done to reduce maternal mortality;
- Antiretroviral therapy is saving lives and must be expanded further;
- Over a quarter of the world's population has gained access to improved sanitation since 1990, yet a billion people still resorted to open defecation;
- 90 % of children in developing regions are attending primary school;
- Old problems such as poverty, inequality, and information gaps persist;
- Cultural rights, communication freedoms, and democracy are missing in MDGs.

Nonetheless, there is wide agreement that the MDGs have played a significant role in addressing poverty, hunger, and disease worldwide. Building on the successes of the MDGs, the SDGs set out an ambitious set of universal "zero"-goals and targets to be achieved worldwide by 2030, including the end of extreme poverty and hunger, gender equality, ensuring quality education, and reducing inequality within and among countries.

6 The SDGs

In September 2015, the 70th session of the United Nations General Assembly in New York set the post-2015 development agenda in the form of 17 SDGs and 169 associated targets. These newly adopted goals are successors to the MDGs. The aim was to overcome the compartmentalization of technical and policy work by promoting integrated approaches to the interconnected economic, social, and environmental challenges confronting the world.

This universal set of 17 goals, 169 targets, and 304 indicators was approved by all UN member states in the UN General Assembly on September 25, 2015. It is meant to be implemented for the next 15 years starting January 2016 (UN 2014, SDSN 2015) (see Annex 2).

The SDGs go further than the MDGs, which aside from having an increased set of agenda to work upon, have more demanding targets (such as the elimination of poverty, instead of reducing its occurrence), as well as closely related and interdependent goals. In general, it is hoped that by 2030, poverty and hunger are eliminated, quality of life is greatly improved, all forms of capital are intact and functioning under ideal climatic situations, and peace and prosperity is shared by all.

Implementing the SDGs follows the Global Partnership for Sustainable Development framework, as outlined in the outcome document of the 3rd International Conference on Financing for Development held in Addis Ababa, Ethiopia, on July 16, 2015 (UNSC 2015). Importantly, the numerous targets indicated in the SDGs require huge capital investments, not just for the SDGs as a whole but also for its individual components. On a global scale, crude estimates have put the cost of providing a social safety net to eradicate extreme poverty at about 66 billion USD a year, while annual investments in improving essential infrastructure could reach seven trillion USD (Ford 2015).

The complexity and interconnectivity of the SDGs requires the mobilization of global knowledge operating across many sectors and regions in order to identify frameworks and systems that would help realize the goals by 2030 (Sachs 2012). Specifically, effective accountability measures need to be in place, such as annual reviews of specific goals, monitoring systems which gather needed information to support and update conducted projects, and follow up statuses of precursor programs essential for meeting a specific SDG as initiated by certain stakeholders (CDP 2015, SDSN 2015). The current framework for achieving and monitoring SDGs which rely on triple bottom line parameters and statistics still needs to be improved through robust conceptual and methodological work (Hak et al. 2016).

7 Communication/Culture and the Sustainable Development Goals (CCSDG): Challenges for a New Generation

On December 17–21, 2015, a five-day conference titled "Communication/Culture and the Sustainable Development Goals (CCSDG): Challenges for a New Generation" provided a forum for about 120 participants from all over the world to explore the complex interplay of communication, culture, and sustainable development in the face of today's global challenges. The conference was coorganized in Chiang Mai by the Southeast Asia Research Center (SEARC) and Department of Media and Communication at City University of Hong Kong, and the Regional Center for Social Science and Sustainable Development (RCSD), Chiang Mai University, in partnership with the Asian Congress of Mass Communication (ACMC), the BGreen Project, the Connect4Climate project of the World Bank, RMIT Melbourne, UNESCO-Bangkok, and Wageningen University.

The concept of this conference was based on a growing realization that communication and culture play a crucial role for achieving sustainability and sustainable solutions, particularly in the current global context, which includes climate change, natural resource degradation, population growth, or increasing fundamentalism. It remains a challenge for researchers to assess the role and importance of communication and culture in the context of the new SDGs, especially for the ASEAN region. The ASEAN region is faced with specific problems, such as land grabbing, dam construction, transboundary management, cross-border migration, and ethnic cultural diversity.

Particularly in regard to culture, it was the first time that the international development agenda has recognized its relevance within the framework of the SDGs, also referred to as the 2030 Agenda for sustainable development, e.g., related to education, the environment, economic growth, or sustainable consumption.

Against this backdrop, the conference centered on different aspects of the complex relationship of communication, culture, and sustainable development and addressed how communication and culture can be understood, defined, and practiced in order to find sustainable solutions for different stakeholders. Current and future challenges and potentials of communication and culture in the context of development were identified and evaluated from different complementary and transdisciplinary perspectives. Key topics included, for example, The transition process from the MDGs to the SDGs, The role of media for sustainable social change, Socio-cultural perspectives on sustainability, The role of indigenous knowledge systems in development issues, Local communities and climate change, Civil society and activism, Religion and sustainable social change, Identity politics and ethnicity, Migration, displacement and exile, The need for digital literacy, Participatory Impact Assessment and learning, The role of public intellectuals, Newsmedia for social change, The role of international organizations, national governments and NGOs, and Western vs Asian perspectives on sustainable social change.

We have selected a number of papers which best summarize the rich diversity offered during the conference and nicely highlight and discuss the crucial challenges for the future.

Madhushala Senaratne's chapter is set against the global discussions and debates surrounding the transition from the MDGs to the new SDGs, and the broad question, what new buzzwords will define development in the years to come. Within this setting, the specific focus of the chapter lies in the buzzwords of "inequality" and "sustainability," recognized as the two most urgent needs of our time, while attention is also placed on the buzzword "poverty" which remains a key connecting element. Senaratne asks three questions: How is inequality and sustainability represented in the texts analyzed?, What new forms of knowledge have emerged about the two buzzwords?, and What buzzwords are set to dominate the new development agenda from 2015 onwards?

The concept of mindfulness was discussed in-depth during the panel discussion "*Mindfulness, Religion and Social Communication*, which explored the predominantly Buddhist tradition and its role in sustainable communication techniques, at the same time as acknowledging the presence of mindfulness characteristics in the belief of 'non-self', which is present in Christianity. According to the panelists, mindfulness can be applied in every aspect of our lives and therefore must be considered in the application of the SDG agenda, in the interpretation of our perceived risks, and in the reduction of suffering. The speakers discussed how mindfulness interplays with the global context of climate change, ongoing terrorist activities and other instances of the human experience.

The role and contribution of religious systems and traditions for the achievement of SDGs, particularly in Asia, is the focus of Anthony LeDuc's chapter. The chapter aims to present the following: (1) stating the reasons why religion is essential to the aspirations of the SDGs in Asia; and (2) explore how the major religions in Asia can contribute to promoting environmental sustainability by providing a framework for (a) assessing the root cause of environmental destruction; (b) envisioning a religious-based approach to how human beings could relate to the natural environment; and (c) presenting how religion promotes harmonious human–nature relationship through a program of self-cultivation and self-transformation. LeDuc argues that the task of achieving the SDGs in Asia is tied with the concern of religion and progress depends greatly on the improved state of human moral and spiritual well-being that religion aims to promote.

Culture, already established as a complex term inclusive of nuanced social dynamics, can be explored from the perspective that culture and identity is what is inherited at birth. In this complex dynamic, each unique culture's preservation is fundamental in the 2030 development agenda, at the same time acknowledging that change is inevitable in development, and therefore, total resistance to it is counterproductive. Within this context, the research of Danesto B. Anacio described a transition which has narratives relevant when considering the achievement of the SDGs. The study, which examined the observed changes to the traditional culture of the Philippine Sagada society, following the arrival of American missionaries in the early 1900s, noted that there were, indeed, rapid changes. Anacio found that some of these changes could be observed in the architecture, the farming practices and the labor systems that are now employed in the region and were not always in line with sustainable practices that were once prioritized. According to Danesto Anacio, while the people of Sagada were welcoming the arrival of the Americans, in hindsight, some of the culture transitions have not always been for the betterment of the community.

Over the last decade, there has been an escalating trend among businesses worldwide to engage in discussions on corporate sustainability and to integrate the SDGs into their business initiatives. The chapter by Kamala Vainy Pillai, Pavel Slutsky, Katharina Wolf, Gaelle Duthler, and Inka Stever presents the findings of an international research project, titled *"Companies' Accountability in Sustainability: A Comparative Analysis of SDGs in five Countries."* The chapter explains how a number of factors (e.g., intense economic competition, or growing stakeholder awareness) have led to a dramatic increase of corporate sustainability (CS) reporting in Asia over the past years. Comparing five countries from emerging markets and developed nations, the chapter focuses on how companies integrate SDGs in their business activities, on their SDG goals and objectives, priorities, and impact measurements. Albeit there is a global trend for businesses to place more emphasis on sustainability, the chapter also shows that companies essentially act as practical agents with varying ethical motivation when engaging in sustainable practices. Image reputation, good government relations, and strategic long-term risk management are among the main motivations of companies for strategically integrating sustainability into their agenda.

Communication and *culture* are both keys to sustainable development, at the same time as being development goals in themselves. To date, development has mostly focused on poverty and education, but the rapid advancement of information and communication technologies (ICTs) is changing that. People can now communicate any time and any place, catalyzing a wider array of opportunities to the development sector. The world today is interconnected and interdependent. By promoting the free flow of ideas, as is the case in UNESCO's mandate, a truly transformative environment can be enabled through the advancement of communication. Disadvantaged groups can now actively participate in their own community's development.

In the regional context, the ASEAN Economic Community (AEC) will hopefully transform Southeast Asia and have a tremendous impact on the 6.8 billion people who call it home, as a unified market and production base fosters a globally competitive economic region. Media and journalists will play an important role in this process, and development workers and civil society, in general, need to ensure that ASEAN meets regional priorities and that its historical experiences are taken into account.

In order to maximize the efficacy of ICTs and the media within the region, development discourse has stated that all members of ASEAN should call on their member states to report on their status of ICT in education and emphasize the importance of ICTs for sharing best practices and experiences within the ASEAN. Pornpun Prajaknate's research is important in this regard. After presenting regional rankings, including the fact that Laos is the only country in ASEAN that has not got an ICT Policy, while Brunei Darussalam, Malaysia, and Singapore have all made the most progress in terms of information and communication technologies, Pornpun Prajaknate names implications in overcoming the divide, including the fact that rural areas are still very much lagging behind the urban areas.

Maya Velmuradova's study can be seen as a response to certain researchers' suggestions of the need to consider communication for development and social change as a technological rather than societal problem. In her multisite study, she investigated the perceived value of use (PVU) concept and modeled the mechanism on small- and medium-sized enterprises (SME) in Central Asia, with the intention to replicate the process in the African context.

G. Arockiasamy, Mallika Vijayakumar, Sujeevan Kumar, and J.B. Anna Asheervadham Mary try to answer the basic question whether poverty can be eradicated through international goals. Building on the thinking of Freire and Sen, they argue that "empowerment" should be used as the guiding principle and that a people's empowerment can be achieved only through peoples' awareness and participation. Their conclusion is, once again, that development is about the development of people and the transformation of society.

Acknowledgments In addition to the organizations and institutions that assisted in the organization of the conference, we wish to explicitly thank Gwang-Jo Kim, director, and Misako Ito, adviser in Communication and Information, at UNESCO's Office in Bangkok; Kanchana Kulpisithicharoen, who coordinated the work of the local organizing committee; and the members of the scientific committee: Fackson Banda (UNESCO, Paris, France), Prof. em. Royal Colle (Cornell University, USA), Dr. Maria Francesch-Huidobro (City University of Hong Kong), Ms Lucia Grenna (C4C, World Bank, Washington DC, USA), Prof. Rico Lie (Wageningen University, Netherlands), Dr. Patchanee Malikhao (Fecund Communication, Hong Kong), Prof. Kiran Prasad (Sri Padmavati Mahila University, Tirupati, India), Prof. Rachel Khan (University of the Philippines, Diliman, Philippines), Prof. Marianne Sison (RMIT Melbourne, Australia), Dr. Song Shi (McGill University, Montreal, Canada), Prof. Jo Tacchi (RMIT Melbourne, Australia), Prof. Chayan Vaddhanaphuti (Chiang Mai University, Thailand), Prof. Karin Wilkins (University of Texas at Austin, USA), and Prof. Loes Witteveen (Wageningen University, Netherlands). We also wish to acknowledge and thank Ms. Jayanthie Krishnan, publishing editor, Mr. Vishal Daryanomel, editorial assistant, and Mr. Naga Kumar, project coordinator, at Springer Nature in Singapore, for their support and assistance.

Annex 1: The Millennium Development Goals (2000–2015)

At the Millennium Summit in September 2000 the largest gathering of world leaders in history adopted the UN Millennium Declaration, committing their nations to a new global partnership to reduce extreme poverty and setting out a series of time-bound targets, with a deadline of 2015, that have become known as the Millennium Development Goals.

The Millennium Development Goals (MDGs) are the world's time-bound and quantified targets for addressing extreme poverty in its many dimensions—income poverty, hunger, disease, lack of adequate shelter, and exclusion—while promoting gender equality, education, and environmental sustainability. They are also basic human rights—the rights of each person on the planet to health, education, shelter, and security.

Goal 1: Eradicate Extreme Hunger and Poverty

Goal 2: Achieve Universal Primary Education

Goal 3: Promote Gender Equality and Empower Women

Goal 4: Reduce Child Mortality

Goal 5: Improve Maternal Health

Goal 6: Combat HIV/AIDS, Malaria and other diseases

Goal 7: Ensure Environmental Sustainability

Goal 8: Develop a Global Partnership for Development

The world has made significant progress in achieving many of the Goals. Between 1990 and 2002 average overall incomes increased by approximately 21 %. The number of people in extreme poverty declined by an estimated 130 million. Child mortality rates fell from 103 deaths per 1000 live births a year to 88. Life expectancy rose from 63 years to nearly 65 years. An additional 8 % of the developing world's people received access to water. And an additional 15 % acquired access to improved sanitation services.

But progress has been far from uniform across the world—or across the Goals. There are huge disparities across and within countries. Within countries, poverty is greatest for rural areas, though urban poverty is also extensive, growing, and underreported by traditional indicators.

Sub-Saharan Africa is the epicenter of crisis, with continuing food insecurity, a rise of extreme poverty, stunningly high child and maternal mortality, and large numbers of people living in slums, and a widespread shortfall for most of the MDGs. Asia is the region with the fastest progress, but even there hundreds of millions of people remain in extreme poverty, and even fast-growing countries fail to achieve some of the non-income Goals. Other regions have mixed records, notably Latin America, the transition economies, and the Middle East and North Africa, often with slow or no progress on some of the Goals and persistent inequalities undermining progress on others.

The final MDG Report found that the 15-year effort has produced the most successful anti-poverty movement in history:

- Since 1990, the number of people living in **extreme poverty** has declined by more than half.
- The proportion of **undernourished people** in the developing regions has fallen by almost half.
- The **primary school enrolment rate** in the developing regions has reached 91 percent, and many **more girls are now in school** compared to 15 years ago.

- Remarkable gains have also been made in the **fight against HIV/AIDS, malaria and tuberculosis**.
- The **under-five mortality** rate has declined by more than half, and **maternal mortality** is down 45 % worldwide.
- The target of halving the proportion of people who lack access to **improved sources of water** was also met.

The concerted efforts of national governments, the international community, civil society and the private sector have helped expand hope and opportunity for people around the world.

Yet the job is unfinished for millions of people—we need to go the last mile on ending hunger, achieving full gender equality, improving health services and getting every child into school. Now we must shift the world onto a sustainable path.

Sources:

http://www.unmillenniumproject.org/goals/
http://www.undp.org/content/undp/en/home/sdgoverview/mdg_goals.html

Annex 2: The Sustainable Development Goals (2015–2030)

At the United Nations Sustainable Development Summit on 25 September 2015, world leaders adopted the 2030 Agenda for Sustainable Development, which includes a set of 17 Sustainable Development Goals (SDGs) to end poverty, fight inequality and injustice, and tackle climate change by 2030.

The Sustainable Development Goals build on the Millennium Development Goals (MDGs), eight anti-poverty targets that the world committed to achieving by 2015. Enormous progress has been made on the MDGs, showing the value of a unifying agenda underpinned by goals and targets. Despite this success, the indignity of poverty has not been ended for all.

The new SDGs, and the broader sustainability agenda, go much further than the MDGs, addressing the root causes of poverty and the universal need for development that works for all people.

Liz Ford of The Guardian summarizes the SDGs as follows: "Sustainable development goals: all you need to know"

http://www.theguardian.com/global-development/2015/jan/19/sustainable-development-goals-united-nations

Are Governments Happy about the Proposed 17 Goals?

The majority seem to be, but a handful of member states, including the UK and Japan, aren't so keen. Some countries feel that an agenda consisting of 17 goals is too unwieldy to implement or sell to the public, and would prefer a narrower brief. Or so they say. Some believe the underlying reason is to get rid of some of the more uncomfortable goals, such as those relating to the environment. Britain's prime minister, David Cameron, has publicly said he wants 12 goals at the most, preferably 10. It's not clear, though, which goals the UK government would like taken out if they had the choice.

Amina Mohammed, the UN secretary general's special adviser on post-2015 development planning, said it had been a hard fight to get the number of goals down to 17, so there would be strong resistance to reducing them further.

Some NGOs also believe there are too many goals, but there is a general consensus that it is better to have 17 goals that include targets on women's empowerment, good governance, and peace and security, for example, than fewer goals that don't address these issues.

How Will the Goals be Measured?

The indicators are still being thrashed out by an expert group. Each indicator is being assessed for its feasibility, suitability and relevance, and roughly two for each target are expected. The indicators are due to be finalised in March 2016.

How Will the Goals be Funded?

That's the trillion-dollar question. Rough calculations from the intergovernmental committee of experts on sustainable development financing have put the cost of providing a social safety net to eradicate extreme poverty at about $66 bn a year, while annual investments in improving infrastructure (water, agriculture, transport, power) could be up to a total of $7 tn globally.

In its 2015 report, the committee said public finance and aid would be central to support the implementation of the SDGs. But it insisted that money generated from the private sector, through tax reforms, and through a crackdown on illicit financial flows and corruption, was also vital.

A major conference on financing for the SDGs, held in the Ethiopian capital Addis Ababa in July 2015, failed to ease concerns that there will not be enough cash to meet the aspirational nature of the goals. The UN said the Addis Ababa action agenda (AAAA for short) contained "bold measures to overhaul global finance practices and generate investment" for tackling the challenges of sustainable development. It included a recommitment to the UN target on aid spending—0.7 % of GNI—set more than 40 years ago and pledges to collect more taxes and fight tax evasion. But civil society groups were less impressed, saying the summit had failed to produce new money to fund the goals, or offer ways to transform the international finance system. Calls for a new international tax body fell on deaf ears.

Sources:

http://www.undp.org/content/undp/en/home/sdgoverview/post-2015-development-agenda.html
http://www.un.org/ga/search/view_doc.asp?symbol=A/RES/70/1&Lang=E
http://www.theguardian.com/global-development/2015/jan/19/sustainable-development-goals-united-nations

References

Agyeman, J. (2013). *Introducing just sustainabilities. Policy, planning and practice*. London: ZED Books.

Avery, G., & Bergsteiner, H. (Eds.). (2016). *Sufficiency thinking. Thailand's gift to an unsustainable world*. Sydney: Allen & Unwin.

Alston, P. (2005). Ships passing in the night: The current state of the human rights and development debate seen through the lens of the millennium development goals. *Human Rights Quarterly, 27*(3), 755–756.

Ban Ki Moon (2016) http://www.unmultimedia.org/radio/english/2016/07/restore-peoples-trust-in-global-economy-by-implementing-sdgs-urges-ban/#.V5xluY78EY0

Besette, G. (2004). *Involving the community: A guide to participatory development communication*. Penang: Southbound/IDRC.

Blewitt, J. (Ed.). (2008). *Community, empowerment and sustainable development*. Devon: Green Books.

Brown, K. (2016). *Resilience, development and global change*. London: Routledge.

Byrne, Elaine, Nicholson, Brian, & Salem, Fadi. (2011). Information communication technologies and the millennium development goals. *Information Technology for Development, 17*(1), 1–3.

CCSDG (2015) *Final Report: Communication/culture and the sustainable development goals (CCSDG): Challenges for a new generation*. Thailand: Chiang Mai University, 17–21 December 2015.

Chambers, R. (2008). *Revolutions in development inquiry*. London: Earthscan.

Easterly, W. (2006) The White Man's burden. Why the West's efforts to aid the rest have done so much ill and so little good. New York: Penguin Books.

Espinosa, A., & Walker, J. (2011). *A complexity approach to sustainability. Theory and application*. London: Imperial College Press.

Flint, W. (2007), Sustainability Manifesto. *Exploring sustainability: Getting inside the concept.* http://www.eeeee.net/sd_manifesto.htm Accessed on July 21, 2015.

Ford, L. (2015). Sustainable development goals: All you need to know. *The Guardian*. Retrieved from: http://www.theguardian.com/global-development/2015/jan/19/sustainable-development-goals-united-nations

Freire, P. (1983). *Pedagogy of the oppressed*. New York: Continuum.

Goodall, C. (2012). *Sustainability. All that matters*. London: Hodder & Stoughton (Winner of the Clarion Award for non-fiction).

Grossman, N. (Ed.). (2015). *Thailand's sustainable development sourcebook*. Bangkok: Editions Didier Millet (EDM).

Hak, T. S. J., & Moldan, B. (2016). Sustainable development goals: A need for relevant indicators. *Ecological Indicators, 60*, 565–573.

Hilty, L., & Aebischer, B. (Eds.). (2015). *ICT innovations for sustainability*. Heidelberg: Springer.

Hopper, P. (2007). *Understanding cultural globalization*. Cambridge, UK: Polity.

Hopkins, R. (2008). *The transition handbook: From oil dependency to local resilience*. UK: Green Books.

Kanninen, Tapio. (2013). *Crisis of global sustainability*. London: Routledge.

Lennie, J., & Tacchi, J. (2013). *Evaluating communication for development. A framework for social change*. London: Routledge Earthscan.

Lesca, N. (Ed.). (2011). *Environmental scanning and sustainable development*. Hoboken NJ: Wiley.

Lie, R. (2003). *Spaces of intercultural communication: An interdisciplinary introduction to communication, culture, and globalizing/localizing identities*. Cresskill: Hampton.

Maclean, K. (2015). *Cultural hybridity and the environment. Strategies to celebrate local and Indigenous knowledge*. Heidelberg: Springer.

Malikhao, P. (2016). *Effective health communication for sustainable development*. New York: Nova Publishers.

Mansell, R., & Raboy, M. (eds.) (2014). *The handbook of global media and communication policy*. Chicester: Wiley Blackwell.

Manyozo, L. (2012). *Media, communication and development. Three approaches*. New Delhi: Sage.

Manzini, Ezio, & Jegou, F. (2003). *Sustainable everyday. Scenarios of urban life*. Edizioni Ambiente: Milan.

Max-Neef, M. (1991). *Human scale development. Conception, application and further reflections*. New York: The Apex Press.
McAnany, E. G. (2012). *Saving the world: A brief history of communication for development and social change*. Champaign: University of Illinois Press.
McMichael, P. (2008). *Development and social change: A global perspective*. Los Angeles: Pine Forge Press.
Nederveen Pieterse, J. (2010). *Development theory* (2nd ed.). Los Angeles: Sage.
Papa, M. J., Singhal, A., & Papa, W. (2006). *Organizing for social change. A dialectic journey of theory and Praxis*. New Delhi: Sage Publications.
Payutto, P. (1998). *Sustainable development*. Bangkok: Buddhadham Foundation.
Rockefeller Foundation (2013). 100 Resilient cities. http://www.100resilientcities.org/#/-_/. Accessed March 29, 2016.
Sachs, J. D. (2012). From millennium development goals to sustainable development goals. *Lancet, 379*(9832), 2206–2211.
Schmidpeter, R., Lu, H., Stehr, C., & Huang, H. (Eds.). (2015). *Sustainable development and CSR in China. A multi-perspective approach*. Heidelberg: Springer.
Scholte, J. (2005). *Globalization: A critical introduction*. New York: Palgrave.
SDSN (2015). Indicators and a monitoring framework for the sustainable development goals. Launching a data revolution for the SDGs. In: *A report by the Leadership Council of the Sustainable Development Solutions Network*. Revised working draft (June 22, 2015).
Sen, A. (2004). Cultural liberty and human development. In S. Fukuda-Parr (Ed.), *Human development report: Cultural liberty in today's diverse World*. New York: United Nations Development Programme.
Servaes, J. (1999). *Communication for development: One World, multiple cultures*. Creskill, NJ: Hampton.
Servaes, J. (ed.) (2007). Communication for development. Making a difference—A WCCD background study. In *World Congress on Communication for Development: Lessons, Challenges and the Way Forward* (pp. 209–292). Washington, DC: World Bank.
Servaes, J. (2007). Harnessing the UN system into a common approach on communication for development. *The International Communication Gazette*, *69*, 6, 483–507 (http://gaz.sagepub.com)
Servaes, J. (Ed.). (2008). *Communication for development and social change*. London: Sage.
Servaes, J. (2011). Social change. *Oxford bibliographies online* (OBO) (58 pp). New York: Oxford University Press http://www.oxfordbibliographiesonline.com/display/id/obo-9780199756841-0063
Servaes, J. (Ed.). (2013). *Sustainability, participation and culture in communication. Theory and Praxis*. Bristol-Chicago: Intellect-University of Chicago Press.
Servaes, J. (Ed.). (2014). *Technological determinism and social change. Communication in a Tech-Mad World*. Lanham: Lexington Books.
Servaes, J., & Lie, R. (2014). New Challenges for communication for sustainable development and social change. A review essay. *Journal of Multicultural Discourses*. (doi:10.1080/17447143.2014.982655)
Servaes, J., & Malikhao, P. (2007). Communication and sustainable development. *FAO, Communication and sustainable development: Selected papers* (pp. 1–38). Rome: FAO.
Servaes, J., & Malikhao, P. (2014). Communication for development and social change: Three development paradigms, two communication models, many applications and approaches 发展及社会变革传播学:三种发展理论,两种传播学模式,众多实践和方法 (Dr. Song Shi, trans.). In J. Hong (Ed.), *New trends in communication studies* (Vol. 1, pp. 20–45). Beijing: Tsinghua University Press (in Chinese) http://www.tup.com.cn http://www.wqbook.com
Jan, Servaes, & Oyedemi, Toks (Eds.). (2016). *Social inequalities, media and communication. Theory and roots*. Lanham: Lexington.
Servaes, J., Jacobson, T., & White, S. (Eds.). (1996). *Participatory communication for social change*. Sage: New Delhi. 287 pp.

Sivaraksa, S. (2009). *Rediscovering spiritual value. Alternative to consumerism from a Siamese Buddhist perspective*. Bangkok: Sathirakoses-Nagapradipa Foundation.
Sparks, C. (2007). *Globalization, development and the mass media*. London: Sage.
Sparks, J., & Rutkowski, D. (2016). Exploring project sustainability: Using a multi perspectival, multidimensional approach to frame design. *Development in Practice, 26*(3), 308–320.
The Economist. (2007, 7 July). The eight commandments, *The Economist*, London, 26–29.
UN (2013). A new global partnership: Eradicate poverty and transform economies through sustainable development. *The Report of the High-Level Panel of Eminent Persons on the Post-2015 Development Agenda*. New York: United Nations.
UN (2014). *Millennium development goals report 2014*. New York: United Nations.
UNDP (2006) Communication for Empowerment: developing media strategies in support of vulnerable groups. (Oslo, Norway) www.worldbank.org/wbi/news/docs/Severinoonfreepress.htm
UNDP (2010). Human Development Report 2010. *The real wealth of nations: Pathways to human development*. New York: United Nations Development Programme.
UNDP (2012) *Mobile Technologies and empowerment: Enhancing human development through participation and innovation*. Available at: http://undpegov.org
UNDP (2014) *Human development report: Sustaining human progress: Reducing vulnerabilities and building resilience*.
UNDP (2015) *Human development report: Rethinking work for human development*.
United Nations Secretary-General's High-level Panel on Global Sustainability. (2012). *Resilient people, resilient planet: A future worth choosing*. NY: UN. ISBN (e978-92-1-055304-9).
UNSC (2015). *Technical report by the Bureau of the United Nations Statistical Commission on the process of the development of an indicator framework for the goals and targets of the post-2015 development agenda* (Working draft). Retrieved from: https://sustainabledevelopment.un.org/content/documents/6754Technical%20report%20of%20the%20UNSC%20Bureau%20%28final%29.pdf
Van Norren, DDorine, E. (2012). The wheel of development: The millennium development goals as a communication and development tool. *Third World Quarterly, 33*(5), 825–836.
Vandemoortele, Jan. (2011a). If not the millennium development goals, then what? *Third World Quarterly, 32*(1), 9–25.
Vandemoortele, Jan. (2011b). The MDG story: Intention denied. *Development and Change, 42*(1), 1–21.
Waisbord, S. (2006). "Where is communication in the Millennium Development Goals?". *Media Development,* LIII (3), 3–6
Whaley, D., Weaver, L., & Born, P. (2010). *Approaches to measuring: Community change indicators*. Waterloo ON: Tamarack. www.tamarackcommunity.ca
Wilkins, K., Tufte, T., & Obregon, R. (Eds.). (2014). *The handbook of development communication and social change*. New York: Wiley-Blackwell.
World Commission on Environment and Development (1987). Our common future. Published as Annex to General Assembly document A/42/427, Development and International Co-operation: Environment. New York NY: UN.
World Economic Forum (2012). Global risks 2012 (7th edition). In *An initiative of the risk response network.* Geneva: WEF.
Zolli, A., & Healy, M. (2012). *Resilience: Why things bounce back.* New York, NY: Free Press.

The Transition from MDGs to SDGs: Rethinking Buzzwords

Madhushala Senaratne

Abstract While, much like its predecessor, 'eradicating poverty' remains the central and overarching narrative of the new development agenda, it looks beyond, affirming renewed and broader emphasis on 'inequality' and 'sustainable development', and at the same time, drawing attention to a set of 'transformative' goals. With the adoption of the new sustainable development agenda that is set to frame development thinking, practice and actions in the next 15 years, this chapter draws attention to selected keywords or buzzwords, specifically that of 'inequality' and 'sustainable development', asking, at a primary level, what do these buzzwords signify? Yet, the chapter also aims to look beyond the conventional. While the United Nations texts provide the primary point of analysis, this chapter is also strongly influenced by the challenge to include fictional representations of development within wider forms of development knowledge, given their ability to offer new insight and perspectives into development issues. Building on this challenge, at the secondary level, the chapter compares and contrasts the findings of policy texts with those of popular media in their representation of inequality and issues of sustainability. The chapter concludes by identifying what knowledge is revealed about 'inequality' and 'sustainable development', while emphasising that 'eradicating poverty' remains a key connecting and compelling buzzwords. It also highlights 'transformation' as an emerging buzzword, arguing that the term remains open to interpretation.

Keywords Millennium development goals · Sustainable development goals · Transition from the MDGs to the SDGs · Sustainable development · Fiction · Sri Lanka · Policy and popular communication

M. Senaratne (✉)
University of Sussex, Brighton, UK
e-mail: H.M.Senaratne@sussex.ac.uk

J. Servaes (ed.), *Sustainable Development Goals in the Asian Context*,
Communication, Culture and Change in Asia 2,
DOI 10.1007/978-981-10-2815-1_2

1 Introduction

In September 2015, the United Nations General Assembly adopted the new sustainable development framework, *Transforming Our World: the 2030 Agenda for Sustainable Development*, an agenda that will guide development policy, thinking and practice over the next 15 years. Building on the plot set forth through its predecessor, the *Millennium Declaration*, the agenda has, as its central narrative, 'eradicating poverty in all its forms and dimensions' (United Nations General Assembly 2015: 1). Embedded in this overarching narrative is renewed and broader emphasis on the notions of 'inequality' and 'sustainable development', while the agenda also introduces a set of 17 goals and 169 targets that are 'people-centred', 'integrated', 'universal' and 'transformative'. Drawing on the existing literature, this chapter regards these key terms and phrases of the new development agenda, such as 'eradicating poverty', 'inequality', 'sustainable development' and 'transformative', as buzzwords. It primarily asks, with the adoption of the new development agenda, what do these buzzwords signify—how do they represent ideas and knowledge of development? Yet, looking beyond the agendas and framework that influences policy, there is a secondary question that the chapter asks taking forth the challenge placed by Lewis et al. (Lewis 2014a, b) to consider wider representations of development to gain alternative insight into today's key development challenges. As such, it asks, how do these buzzwords compare with their alternative understandings—understandings that ordinary folk may derive of them?

Sachs (2010a: xvi) once contended, 'development is much more than just a socio-economic endeavour; it is a perception which models reality, a myth which comforts societies, and a fantasy which unleashes passions'. Building on this premise, Cornwall (2010: 1) regards these models, myths and fantasies as being 'sustained by development's buzzwords'. Buzzwords define what is in vogue and produce certain frames of knowledge and understanding of 'development' (Cornwall 2010). They are passwords for funding and influence, and justify interventions, yet, remain ambiguous, open to diverse interpretations and are often taken for granted (Cornwall 2010). Buzzwords are such that, amidst delusion, disappointment and disaster, they continue to dominate global and national policies as well as the thinking and language of grass-roots organisations (Sachs 2010a: xv). And it is here that development's mystery lies.

This chapter is placed against the broader debates and discussions surrounding the transition from the millennium development goals (MDGs) to the sustainable development goals (SDGs) and is particularly influenced by two pieces of edited volumes, the *Development Dictionary* and *Deconstructing Development Discourse: Buzzwords and Fuzzwords*, which deconstruct selected buzzwords that make up today's development lexicon. It argues that, against the backdrop of the adoption of the new sustainable development framework, it is perhaps the time to set about the task of unravelling the mystery of development's buzzwords once again, while at the same time, looking beyond the conventional meanings of development policy to their broader understandings in popular representations.

In the 2000s, the MDGs became 'a rallying force for development' with its promising and unifying narrative of development through poverty eradication (Kharas and Zhang 2014). Since then, among the significant progresses made in achieving the targets set out through the MDG framework were efforts in improving the lives and conditions of those living in extreme poverty. The *Millennium Development Goals Report 2015* notes that globally, the number of those living in extreme poverty "declined by more than half, from 1.9 billion in 1990 to 836 million in 2015" (United Nations 2015). Despite these achievements, millions across the world are 'being left behind, especially the poorest and those disadvantaged because of their sex, age, disability, ethnicity or geographic location' (United Nations 2015: 8). Inequality—gender inequality and widening gaps between rich and poor households, rural and urban—as well as climate change and environment degradation, and conflict are identified as being among the biggest threats to human development (United Nations 2015). Inequality is increasingly seen as hindering the well-being of nations and people, increasing instability, and political and social tensions (UNDP 2013), while emphasis has also been laid on the need to promote sustainable development that takes into consideration the needs of both the people and the planet (United Nations 2014) (see also Servaes and Oyedemi 2016a, b).

It is against this justification that this chapter seeks to draw understandings of the common phrases of 'inequality' and 'sustainable development', while examining 'poverty eradication' as a key connecting and compelling buzzword that continues to be central in development policy narratives, and identifying 'transformative' as an emerging keyword of today's changing development lexicon.

In examining these buzzwords, this chapter considers selected texts and adopts a two-pronged approach. At a primary level, this chapter recognises that policy documents provide key insight and information into development processes, as well as spell out the development plans and actions. As such, findings are drawn from key United Nations texts, including the *Road Map towards the implementation of the UN Millennium Declaration (2001)* and *Transforming our world: the 2030 Agenda for Sustainable Development (2015)*. In addition, given that this chapter is derived from an earlier piece of work that was undertaken prior to the adoption of *Transforming our world*, the analysis also draws on *Road to Dignity by 2030: Synthesis report of the Secretary-General on the Post-2015 Agenda (2014)*, which was initially examined.

Yet, how do these findings of 'poverty', 'inequality' and 'sustainable development' fit in with broader understandings of the buzzwords? As such, at a secondary level, I am influenced by the challenge that Lewis et al. (2014a, b) places, to look beyond conventional forms of representations of development and include broader and wider forms of representations in order to gain alternative insight into key development concerns, debates and discussions. Thus, the findings of policy texts are compared and contrasted with those of popular media in their representation of development, specifically, inequality and issues of sustainability.

As such, along with the United Nation texts, the fiction selected for analysis are written by Sri Lankan authors: *Samsara* (2001) and *Sam's Story* (2009). The reason for the selection of Sri Lankan texts is driven partly by my own position as a Sri Lankan and the country's changing development landscape. Sri Lanka has made steady progress in achieving the MDGs through the escalation of a nearly three-decade-old conflict and a post-conflict era. The country's positive MDG performance as compared to other South Asian countries is also significant. For example, in reduction of poverty and hunger levels, and achievements in education and wage employment, Sri Lanka has made noteworthy progress in the region, while in improving maternal health the country's achievements are on par with those of more advanced economies (United Nations Sri Lanka 2015). Yet, as Sri Lanka looks set to embrace the new SDGs, new challenges have emerged, specifically in the areas of inequalities within and between regions, environmental sustainability and issues of reconciliation that question the key notions of sustained and inclusive growth and environmental protection. For many Sri Lankan writers, the country's conflict provided the setting and a central narrative for their work, giving them a platform to express aspects of the struggle that remained unspoken, and explore the conflict through different lenses.

2 Media as Vehicles of Meaning Production

Media and texts act to produce meaning through language (Hall 1997). Language constructs and transmits meaning operating as a 'representational system', where signs and symbols, for example, written words, represent concepts, ideas and feelings about specific events, enabling the audience to read, decode or interpret their meanings (Hall 1997). Semiotics, the study or science of signs, provides a general model of understanding the role of language as vehicles of meaning and representation in society (Hall 1997). Ferdinand de Saussure distinguished between two elements of the 'sign' which were fundamental to the production of meaning (Hall 1997). He identified the actual written word or image as the 'signifier', while the mental concept or idea associated with that word was defined as the 'signified' (Hall 1997). However, signs are arbitrary (Deacon et al. 2007), and the value of any sign derives from its relation to other signs (Chouliaraki and Fairclough 1999).

The texts produced by the UN have been chosen for analysis given their over-arching and universal nature. *The Road map towards the implementation of the United Nations Millennium Declaration (2001) (Henceforth referred to as: Road Map)* is identified given its significance as setting a new normative framework for international development in 2000 (Kharas and Zhang 2014). The 2000 Millennium Summit, which saw the adoption of the Millennium Declaration, and the subsequent MDGs recorded the largest gathering of world leaders at the time (Fukuda-Parr and Hulme 2009). The Summit saw leaders 'committing their nations to a new global partnership to reduce poverty' and achieve the MDGs by 2015 (United Nations 2000). Similarly, the *Road to Dignity by 2030: Synthesis Report of the Secretary-General on the Post-2015 Agenda (2014) (Henceforth referred to as: Road to Dignity)* and *Transforming Our World: The 2030 Agenda for Sustainable Development (2015) (Henceforth referred to as: Transforming Our World)* were selected primarily against their significance as defining the development agenda from 2015 onwards. The texts introduce the new post 2015 development agenda aimed at 'ending poverty, transforming all lives and protecting the planet' (United Nations 2014: 1). While building on the MDG narrative, the SDGs look beyond with the aim of promoting sustained and inclusive economic development and environment protection (United Nations 2014).

Yet, according to Escobar (1995), there is a need for more research on the languages of development at the local level to derive a better understanding of some of the key concepts and modes of operation. While policy and academic literature construct knowledge of development problems in ways that justify policy responses, literary works offer new insight and alternative understanding into development concerns (Lewis et al. 2014a, b). It is against this rationale that findings of the policy texts are contrasted and compared with popular communication.

Samsara (from the short story collection: *In the garden secretly and other stories*) by Jean Arasanayagam was published in 2000. The short story collection received the Sri Lanka State Literary Award in 2001. A collection of seven stories, it highlights the themes of war, rebellion, displacement and dispossession. *Samsara*—the last story of the collection—tells the story of Mudiyanse, who helps the narrator—a public school teacher—clear their garden of wild growth. It is through the narrator's interactions with Mudiyanse that his story emerges. Following the death of his parents, Mudiyanse is cast aside by his family and denied his share of inheritance of land, leaving him destitute as he now makes a meagre living by working in the houses of the estate, carrying out tasks such as cleaning gardens, cutting grass and sometimes helping with construction work. However, his obsession of reclaiming what is rightfully his leads him to be labelled as a madman and cast aside by his own villagers. *Sam's Story* by Elmo Jayawardena was first published in 2001, with the manuscript having received the Gratiaen Prize for Fiction in the same year. The version of the text chosen for analysis is a later edition, published in 2009. *Sam's Story* is narrated through Sam who comes from an impoverished background to work as a houseboy in an upper-middle-class household. The story shifts between Sam's current comfortable circumstances to his impoverished childhood with powerful imagery that gives a voice and a face to the poor (Perera 2002). Using irony and humour, *Sam's Story* draws a stark understanding of the 'rich versus poor' notion, while also highlighting how illiteracy and lack of awareness in rural settings can lead to harmful impacts on the environment and using the ethnic conflict and the country's changing political climate as the backdrop to the story.

3 Changing Buzzwords and the Role of Fiction in Development

3.1 Development: Changing Definitions and Buzzwords

The definition of development remained vague after it emerged as a password in the post-Second World War period (Rist 2010). The 'program of development' presented by Henry Truman in 1949 provides a useful starting point in understanding the meaning of development.

> We must embark on a bold new program for making the benefits of our scientific advances and industrial progress available for the improvement and growth of underdeveloped areas (Esteva 2010: 1).

There is echo of President Truman's sentiments in crucial United Nations documents of the time, which aimed to design 'concrete policies and measures for the economic development of underdeveloped countries' (Escobar 1995: 4). Such representations of development are hegemonic and ethnocentric (Escobar 1995; Esteva 2010).

According to Servaes and Malikhao (2008: 159), 'the central problem of development was thought to revolve around the questions of 'bridging the gap' and 'catching up' by means of imitation processes between traditional and modern sectors…' These phrases, 'catching up' and 'bridging the gap', capture the essence of economic development equality (Lummis 2010). Lummis (2010) provides a useful introduction into 'equality' while highlighting its vagueness and multiple significations. Steadily accelerated economic growth can lead to equality or towards reducing inequality (Lummis 2010), while inequality is also reflected in poverty and insecurity (Standing 2010).

While inequality thus remained a key buzzword, words such as 'interdependence' 'integration', 'global' and 'unified' also gained prominence in the development lexicon (Servaes and Malikhao 2008). Yet, the move towards more local responses to development, the failure of top-down development projects and the increased attention paid towards economic and technical factors as leading to social and ecological problems led to the rise of the 'sustainable development' or the 'sustainability' narrative (Escobar 1995). This narrative was further reinforced through the 1992 document, *Our Common Future* also known as the *Brundtland Report*, which captured the essence of the narrative through its definition of sustainable development as development that meets the needs of the present without compromising the ability of future generations to meet their own needs (Sachs 2010b).

3.2 The Role of Fiction in Development

According to Pawling (1984:4), 'popular fiction reflects social meanings and […] intervenes in the life of society by organising and interpreting experiences which have previously been subjected only to partial reflection'. Within the overall notion of popular fiction, Nussbaum (1995) identifies novels as particularly significant in public reasoning. The reason for this, Nussbaum (1995) argues, lies in its ability to foster 'sympathetic imaginations'. Novels,

> …construct and speak to an implicit reader who shares with the characters certain hopes, fears and general human concerns, and who for that reason, is able to form bonds of identification and sympathy with them (Nussbaum 1995: 7).

Thus, novels not only represent development issues, but also enact feeling and imaginations, allowing readers to recognise, connect and 'form bonds of sympathy' with the inner lives of others, various human needs, desires and specific social situations, most often with people and situations vastly different to that of the reader (Nussbaum 1995). They portray everyday realities, experiences, and human and social conditions, allowing readers to identify with such realities (Pawling 1984).

These arguments are echoed in the more recent works of Lewis et al. (2014a, b), which formed a key influence of this research. Emphasising that fiction cannot be considered as fact and should be considered alongside products of development, such as policy reports, academic or scholarly writings, they argue, fiction produces valid and alternative knowledge into development processes, while also capturing the humanistic side of development that is often overlooked in policy documents (Lewis et al. 2014a, b). The power of fiction also lies in its accessibility, wide reach and ability to engage readers by humanising issues (Lewis et al. 2014a, b). Thus, with little development of jargon and buzzwords, as will be evident through the argument of this chapter, fiction perhaps offers different ways of understanding of what we may or may not know about 'poverty', 'inequality' and 'sustainable development'.

4 Buzzwords in Development Policy

4.1 Poverty

'Poverty' remains one of the most compelling buzzwords in development policy literature (Cornwall 2010). Yet, *Road Map* highlights a key problem in the signifier and the signified of 'poverty':

> Development and poverty eradication: the millennium development goals
>
> In order to significantly reduce poverty and promote development it is essential to achieve sustained and broad-based economic growth. The MDGs highlight some of the priority areas that must be addressed to eliminate extreme poverty.
>
> (United Nations General Assembly 2001: 18–19)

In addition, a later part of the text makes reference to 'alleviating poverty'.

While emphasis remains on the phrase, 'poverty eradication', its use almost interchangeably with 'reducing' and 'alleviating' poverty is a main problematic. Toye (2007) identifies poverty alleviation, poverty reduction and poverty elimination as carrying three different significations. He regards poverty alleviation as 'temporising' the concern, rather than confronting it, while 'poverty eradication', although carrying the 'right degree of gritty determination and radicalism' has a 'utopian feel about it' (Toye 2007: 47). Therefore, 'poverty reduction' is recognised as the ideal choice of action in improving the conditions of the poor, given the most resolute and realistic of the three (Toye 2007).

Significantly, in the *Road Map*, 'poverty' and its related phrase, 'poverty eradication', are presented in conjunction with other buzzwords, specifically, 'development' and 'economic growth', constructing a strong link between poverty eradication, economic growth and development, with economic growth as essential to eradicating poverty, which in turn would promote development.

This then leads to the question of what is signified by poverty and poverty eradication. Rahnema (2010) argues that in ancient Europe, the poor—or the pauper—was opposed only to the powerful, rather than the rich. On the other hand, there were—and still are—references made to those 'living poorly' or the voluntary poor (Rahnema 2010). More commonly, however, 'poverty' converges an image of 'a kind of generalised lacking, or a state of being without some essential goods and services' (Toye 2014: 45). It signifies the beggar and the street vendor, the homeless, landless, the weak and the hungry (Rahnema 2010). Poverty is also a feature of underdevelopment (Mohanty, as cited in Escobar 1995). There are clues in the text that provide meaning to poverty, specifically its association with those whose income is 'less than one dollar a day (2001: 19), while the *Millennium Declaration* (United Nations General Assembly 2000: 4) speaks of the 'dehumanising conditions of extreme poverty'. Poverty eradication and development then, is, making efforts to 'free [...] fellow men, women and children' from these 'dehumanising conditions' (United Nations General Assembly 2000: 4).

4.2 Inequality

While 'poverty eradication' is thus highlighted, the analysis of the *Road Map* indicates the lack of the use of, 'inequality' and 'equality'. This is significant given that the Millennium Declaration recognises the 'collective responsibility to uphold the principles of [...] equality' (United Nations General Assembly 2000: 1). The primary mention of 'equality' in the *Road Map* is in relation to gender, 'To promote gender equality and the empowerment of women as effective ways to combat poverty, hunger and disease, and to stimulate the development that is truly sustainable (United Nations General Assembly 2001: 24)'. Thus, while equality is a concern, it has come to be used in development more often with gender (Smyth 2010), signifying an empowered woman, with equal rights and opportunities as a man.

Standing (2010) argues that inequality is reflected in poverty and insecurity. Yet, signs are arbitrary, and considering their signified can be subject to history (Hall 1997). Living in Sri Lanka, against the backdrop of a 27-year-old conflict that caused tension between the country's Sinhalese majority and the Tamil minority, the buzzword 'equality' draws a mental image of all segments of the community, that is, both males and females of all ethnic communities of the country, having access to equal opportunities, enjoying the fruits of development equally and working together for peace. This image is perhaps cliché. Yet, equality is present in any belief that people ought to come under the same set of rules (Lummis 2010). Further breaking down the signified of equality, Lummis (2010) goes on to identify two families of meaning of the buzzword: equality as justice or fair treatment, and equality as indicating sameness or homogeneity.

In contrast to the *Road Map*, the new development agenda affords greater emphasis on 'inequality'. *Road to Dignity* states that the new sustainable development agenda must 'address inequalities in all areas, agreeing that no goal or target be considered met unless met for all social and economic groups' (United Nations 2014: 19). This statement is listed as an essential element of the new agenda. In addition, unlike in the *Road Map*, *Transforming Our World* recognises the multiple dimensions of inequality with the use of the signifier in relation to gender, income and opportunity. Further, sustainable development goal 10, reads, 'reduce inequality within and among countries' (United Nations General Assembly 2015).

Such use of the buzzword, and its use in relation to other keywords, is significant in drawing much-needed renewed attention to ensuring 'equality', and the need to address inequalities. However, the analysis observes that such a use of the buzzword in the new development agenda contributes to an alternative understanding of the signified of the term as portrayed by Standing (2010) in his reflection that the primary objective of the welfare state, while some believe should be to promote happiness, is also to promote social justice, income inequality and equal life chances.

4.3 Sustainability

Scoones (2010) regards 'sustainability' as a boundary term, linking environment and economic development concerns, and as one of the most widely used buzzwords in recent decades. Early signifiers of 'sustainability' meant managing forests in order to preserve them on a long-term basis (Scoones 2010) and, as discussed in the literature review, it was the text, Our Common Future. Another signifier of 'sustainability', drawing on Escobar (1995), is associated with the notion that it represents bottom-up development, involving wider community participation and local initiatives in development processes.

Emphasis on 'sustainability' in the *Road Map* is made through the notion of building a framework for a global partnership to accelerate sustainable development in least developed countries and small island developing states. In comparison, along with inequality, sustainability is given similar emphasis in the new development agenda, which highlights the need to 'integrate sustainability in all its activities, mindful of economic, environmental and social impacts' (United Nations 2014: 19). *Transforming our word* contains further related phrases such as sustainable agriculture, sustainable management of water and sustainable industrialisation. Drawing on Scoones (2010), such phrases signify good development. Scoones (2010) observes that the link between the two strands of science and policy through such phrases and pairings of the terms can be a positive force in development and goes on to pose the question, will sustainability become the unifying concept of the twenty-first century, linking different groups? He is confident that 'sustainability and its wider agenda is here to stay' (Scoones 2010: 89).

5 Buzzwords of Popular Fiction

5.1 Poverty, Exclusion and Enforced Differences

In *Samsara*, comparison is drawn between the 'housing estate' where the narrator lives and '*wadiya*' or the hut of Mudiyanse (Arasanayagam 2000). The hut is described as a 'fragile shelter' covered with 'dried coconut fronts' which keeps out the rain and cold air at night (Arasanayagam 2000: 136). Such imagery acts to signify poverty and powerlessness. Yet, the narrator herself lives in a housing estate. In Sri Lanka, the term 'housing estate' refers to a type of public housing, subsidised or low-income housing, provided for government employees of ministries and services, such as public school teachers (Samaratunga 2013). Within the sign of 'housing estate', drawing on Samaratunga (2013), is the signifier, meaning 'economically weaker groups in society'. A mental image of these houses as lowly and of convenience is further formed as it is narrated that they 'stuck like awkward protuberances in the landscape, unlike the houses in the original village, which merged in with the earth' (Arasanayagam 2000: 135). This serves as a metaphor for the lives of the narrator and Mudiyanse, who, portrayed as 'an outsider' and 'scarecrow', depicting a sense of displacement and exclusion within their own community (Arasanayagam 2000). For example, the narrator parallels him to an 'old weathered scarecrow' (Arasanayagam 2000: 148). I argue here that 'scarecrow' has multiple significations. On one hand, it depicts the notion of simplicity, while on the other hand, it carries with it the meaning of powerless to act or speak. The narrator, however, is able to sympathise with Mudiyanse, as she reflects, 'we too felt like outsiders' having being displaced themselves several times and struggling to service on their meagre income (Arasanayagam 2000: 148). By thus sympathising with Mudiyanse, the narrator also draws our attention to the need to feel empathy for the other, highlighting that it is only then that another would understand their condition.

Mudiyanse's obsession with reclaiming his lost inheritance led him to be regarded as a madman and being rejected by his own village and friends (Arasanayagam 2000). Sent away by his own brothers to a mental asylum when his demands got more insistent, Mudiyanse reflects,

> The people began to say "*pissu, pissu* [mad, mad]". I was not mad at all. One day they fastened manacles on my wrists and I was taken away. While I was there, in the asylum, my whole body would tremble when I was subjected to the electric shocks treatment (Arasanayagam 2000: 139).

Similarly, Mudiyanse's treatment as an outside is further evidenced in the village children calling him, 'crocodile', with the narrator stating, 'the animal image had its special connotations of fear and threat' (Arasanayagam 2000, p. 140). Through this animal image, Mudiyanse is also dehumanised, reflecting the dehumanising conditions of poverty that the Millennium Declaration (United Nations General Assembly 2000) speaks of.

What is significant, and alarming here, is that Mudiyanse's status as an outcast and a madman is constructed by society itself—it is a label accorded to him by his own village. Drawing on Escobar (1995), evident here is the notion of powerful groups constructing a certain image of the powerless, as an outsider. At the same time, Pawling (1984: 117) argument is also valid that 'perceptions of snobbery' lead to inequalities.

Much like Mudiyanse who lived in a hut, Sam speaks of his home, 'together in that one-roomed shack, the six of us had done our best to make something for our lives' (Jayawardena 2009: 102).

The readers are taken further in rural life as Sam recollects,

> There isn't much that has changed in our village. The men still dig sand from the river and drink *kasippu* [strong alcohol] in the evening to forget their troubles. The women go to tap rubber from the trees and come home to empty kitchen and try to find things to cook for their nightly meal (Jayawardena 2009: 172)

The text thus provides us insight into Sam's life and an alternative image into the notion of poverty. Such imagery, also helps the reader, as Nussbaum (1995) argues foster one's sympathetic imagination with the reader made to see, hear, think and feel the pain and sufferings of the poor. Sam's innocent description of the poor is powerful raising questions on their conditions of living, their ability to survive, right to decent work and access to basic services.

5.2 *Inequality*

As opposed to the vagueness of 'inequality' in policy texts, the statement below provides an alternative signified.

> Back home in our village, it was very difficult when somebody fell sick. That is why so many died so young. There were no vans [...]; no one to carry us to the doctor [...]. We were very lucky if we got medicine.
>
> When we fell ill seriously we had to find our own way to the nearby town and wait for hours in a long line to see the doctors who worked in the government hospitals [...] At such times, we were worse than dogs.
>
> Bhurus was lucky. He was from the River House. The River House was a rich house. That made all the difference (Jayawardena 2009: 76).

Bhurus is the dog in the River House where Sam serves as a houseboy. Summarising Sam's words, inequality is seen here as the difference between the rich and the poor. It is ironic that while even a dog of a rich house receives care in sickness, the poor have little access to services and face injustice, sometimes treated 'worse than dogs'. The negative impact of such inequalities and poverty is also highlighted, as many villagers died with little access to care. Thus, in Sam's understanding, being rich made all the difference.

5.3 *Sustainable Development*

Sam draws readers' attention to another vital question related to sustainable development, as he recounts how in the village they threw everything, including all dirt, into the river along with their houses were built (Jayawardena 2009). Having moved into the River House, Sam is taken aback and fails to understand why the Madam (the Master's wife) constantly reminds him not to throw dirt into the river by the house and instead collect it in garbage bags (Jayawardena 2009). Sam's confusion is evident, as he throws dirt into the river when the Madam is not there, asking, 'I don't know why she was worried about collecting dirt in bags and not throwing it all in the river [...] She spoke as if the river would mind' (Jayawardena 2009: 13). Sam's statement is innocent, comical and at the same time raises alarming concern, as he leaves us with a question, how do we tell him that the river would mind?

6 Changing Buzzwords

The purpose of using both policy and popular media was to better understand 'the relationship between different accounts and forms of representation within development writing as well as noting the multiplicity of voices and logics' (Lewis et al. 2014a, b: 30). Against this statement, I argue that we have now heard from distinct voices. What links all these voices together, I argue, is the common theme, buzzword or signifier of 'poverty', while its logics and representations provide a point of comparison.

Until now, the two strands of texts of policy and popular media were considered separately. Yet, how do their findings compare and contrast with each other? While 'eradicating poverty' remains central to the new development agenda, given the framework's renewed emphasis on 'inequality' and 'sustainable development', this discussion specifically asks, what type of knowledge and understanding is revealed about these buzzwords that feature prominently in today's development lexicon?

6.1 *Inequality*

The cross-analysis of the texts reveals that inequality remains a contested buzzword, open to multiple significations. Based on the findings of the analysis, I categorise the following types of knowledge revealed about 'inequality':

Inequality resulting from a 'rich versus poor' syndrome, leading to the urgent need to look beyond merely eradicating poverty.

As discussed earlier, Sam makes a powerful assertion that being rich 'made all the difference' (Jayawardena 2009: 76). Making his statement count further—almost as if meaning to cause some embarrassment—the poor is compared to the dog of the rich household. While inequality here is closely linked to poverty, it also raises a haunting alternative that reducing poverty would not merely address the issue, but rather, there is a need to consider the causes and structures that lead to poverty, a matter that has little focus in policy documents. However, it must be stated that, as opposed to the *Road Map* that used inequality mainly in conjunction with gender, *Road to Dignity* is promising in this respect with its emphasis not just on reducing poverty, but also on inequalities, and in all areas of development.

Inequality and 'the perception of snobbery'.

Pawling (1984: 117) regards the 'perception of snobbery as the crucial ethical problem in an unequal society'. Perception is embedded in representation. Representation, for example, leads to the construction of certain perceptions and frames of understanding (Hall 1997). Escobar (1995) and Esteva (2010) perceived 'development' as hegemonic and a Eurocentric ideal. Applying such arguments in a narrower sense, *Samsara* highlights how the notions of hegemony and power are evident within social groups and structures itself. Mudiyanse is portrayed as a madman, fuelling ideas of an outcast and a threatening being (Arasanayagam 2000). Further, Mudiyanse is dehumanised, reduced to an animal, as he is labelled a crocodile. The difference here is enforced and constructed by society itself. Therefore, drawing on Pawling (1984), by contrasting fiction with the findings of policy texts, what emerges is a society, where difference, unhappiness and the dehumanising conditions in which people live can very well be enforced, with those living in extreme poverty, with little knowledge of systems and access to services, are pushed further towards the status of the subaltern in the social structure, leading to inequalities.

Equality as providing equal life chances, development for all and in all sectors.

Transforming our world takes the notion of inequality further in their pledge that 'no one will be left behind' (United Nations General Assembly 2015: 1), and that 'no goal or target be considered met unless met for all social and economic groups' (United Nations 2014: 19). Significantly, the new development agenda goes on to recognise the multiple dimensions of inequality with regard to gender, income and opportunity, while also identifying the need to 'reduce inequality within and among countries' (United Nations General Assembly 2015: 14). Thus, with the new development agenda, a broader notion of reducing inequality is presented, signifying ideas and knowledge of equal life chances, opportunities and access to services for all segments and sectors of society.

6.2 Sustainability

In contrast to the understanding of inequality, what is noticed is the limited attention paid to the notion of sustainability in literary texts. However, based on the findings, the following understandings have been derived of the buzzword:

Sustainability as requiring change, but also transformative.

Both policy texts link the buzzword closely with 'partnership'. While the *Road Map* highlights the need for a global partnership for sustainability development specifically with regard to the least developed small island states (United Nations General Assembly 2001), *Transforming our world* speaks of the need to 'strengthen the means of implementation and revitalise the global partnership for sustainable development' (United Nations General Assembly 2015: 14). It recognises that sustainable development requires transformations; for example, in building partnerships that must be inclusive and participatory involving all stakeholders and accountable to people, which in turn, has the potential to transform societies. Further, the text affirms that sustainable approaches must be adopted in all aspects of development and production, including agriculture, manufacturing, water and sanitation and job creation (United Nations General Assembly 2015) and, drawing on Scoones (2010) sustainability thus becomes crucial for good development. Such emphasis leads to an urgent call for change, and perhaps confirms a further argument raised by Scoones (2010) that sustainability is here to stay.

Sustainability and knowledge—would the rivers mind?

Popular media leads to the construction of an alternative type of knowledge relating to sustainability. For example, Sam recounts how the men in his village still dug sand from the river bed and how most of the village houses situated near the river threw everything—all the dirt—in the river (Jayawardena 2009). Sam continues his habits when he goes to work at the River House and fails to understand why the Master's wife insists that the dirt is collected in garbage bags instead of dumped into the river. Here, one is left to wonder, whether Sam was more aware of the impact of his activities, will his condition and attitude change? Ironically, *Road to Dignity* recognises that human activities are at the centre of posing a threat to sustainable development and points to the harmful nature of such activities (United Nations 2014). As highlighted earlier, the text also speaks of the need for sustainable approaches in all sectors and livelihoods, including managing water resources, waste and chemicals (United Nations 2014). This emphasis is significant; however, taking stride from Sam's situation, we wonder, is what promised in the new development agenda sufficient?

7 Conclusion—From Old to New Buzzwords

Scoones (2010: 160) asks, 'can old buzzwords be reinvigorated and reinvented for new challenges, or does it need discarding, with something else put its place?'

Transformation is defined as the new development agenda's 'watchword' (United Nations 2014: 3). With the buzzword implying 'change', there is emphasises on the need to 'embrace change', in the 'management' of societies, economies and the planet (United Nations 2014). This transformation is to be realised by promoting patterns of growth that are more inclusive, sustained and sustainable, which in turn, will help tackle issues such as climate change, promote effective governance and ensure good development (United Nations 2015). Yet, the signification of 'transformation' or 'transformative' as portrayed in the new agenda is rather ambiguous, as it emphasises on the need for transformations in different aspects, such as rural development, agricultural systems, trade and financial sectors.

Cornwall (2010: 13) argues that it is given 'the very ambiguity of development buzzwords that scope exists for enlarging their application to encompass more transformative agendas'. It is here then, in the very ambiguity of buzzwords, that development's mystery lies. The broad nature of buzzwords holds promise, multiple interpretations and disappointment. But, buzzwords continue to matter in development, specifically for the reason that they continue to draw attention to, frame understandings and produce knowledge about key development concerns and ways of addressing them. The sustainable development narrative with its promise to leave no one behind is appealing. Yet, as the new sustainable development agenda is being put into action, and as the agenda's dominant buzzwords of 'inequality' and 'sustainable development', along with its promise of 'transformation' and the overlying objective of 'eradicating poverty', continues to define and persuade policy, I am particularly drawn to what Sam says—rather, pleads,

> Sometimes I wished they could all become poor, I mean really poor like my family; at least for a short time. Then they would know what this business of being poor was all about (Jayawardena 2009: 159).

References

Arasanayagam, J. (2000). Samsara In *In the garden secretly and other stories.* New Delhi: Penguin.

Chouliaraki, L., & Fairclough, N. (1999). *Discourse in late modernity: Rethinking critical discourse analysis.* Edinburgh: Edinburgh University Press.

Cornwall, A. (2010). Introductory overview—buzzwords and fuzzwords: deconstructing development discourse. In A. Cornwall & D. Eade (Eds.), *Deconstructing development discourse: Buzzwords and fuzzwords.* Rugby: Practical Action Publishing in association with Oxfam GB, 1–18. [Online] http://www.guystanding.com/files/documents/Deconstructing-development-buzzwords.pdf. Accessed 28 April 2016.

Deacon, D., Pickering, M., Golding, P., & Murdock, G. (Eds.). (2007). *Researching communications: A practical guide to methods in media and cultural analysis.* London: Hodder Arnold.

Escobar, A. (1995). *Encountering development: The making and unmaking of the third world.* Princeton, NJ: Princeton University Press.

Esteva, G. (2010). Development. In W. Sachs (Ed.), *The development dictionary: A guide to knowledge as power* (2nd ed., pp. 1–23). London: Zed. [Online] http://www.citizens-international.org/ci2012/http:/www.citizens-international.org/ci2012/wp-content/uploads/2012/12/Development-Dictionary.pdf. Accessed 23 April 2016.

Fukuda-Parr, S., & Hulme, D. (2009). International norm dynamics and 'the end of poverty': Understanding the millennium development goals (MDGs). *BWPI Working Chapter* 96. [Online] http://www.eadi.org/typo3/fileadmin/MDG_2015_Publications/fukuda-parr_and_hulme_2009_international_norm_dynamics.pdf. Accessed 24 April 2016.

Hall, S. (1997). *Representation: Cultural representations and signifying practices.* London: Sage.

Jayawardena, E. (2009). *Sam's story* (2nd ed.). New Delhi: Penguin.

Kharas, H., & Zhang, C. (2014). New agenda, new narrative: What happens after 2015? *SAIS Review of International Affairs,* 34(2), 25–35. [Online] http://muse.jhu.edu/journals/sais_review/v034/34.2.kharas.pdf. Accessed 24 April 2016.

Lewis, D., Rodgers, D., & Woolcock, M. (2014a). The fiction of development: Literary representation as a source of authoritative knowledge. In D. Lewis, D. Rodgers, & M. Woolcock (Eds.), *Popular representations of development: Insights from novels, films, television and social media* (pp. 3–16). Abingdon: Routledge.

Lewis, D., Rodgers, D., & Woolcock, M. (2014b). Introduction. In D. Lewis, D. Rodgers, & M. Woolcock (Eds.), *Popular representations of development: Insights from novels, films, television and social media* (pp. 3–16). Abingdon: Routledge.

Lummis, A. T. (2010). American Muslim Women: Negotiating Race, Class, and Gender within the Ummah. *Sociology of Religion,* 71(4), 488–490.

Nussbaum, M. (1995). *Poetic justice: The literary imagination and public life*. Boston: Beacon Press.

Pawling, C. (Ed.). (1984). *Popular fiction and social change*. London: Macmillan Press.

Perera, S. W. (2002). Sri Lanka. *The Journal of Commonwealth Literature, 37*(3), 195–205. [Online] doi:10.1177/002198940203700308. Accessed 24 April 2016.

Rahnema, M. (2010). Poverty. In W. Sachs (Ed.), *The development dictionary: A guide to knowledge as power* (2nd ed., pp. 1–23). London: Zed. [Online] http://www.citizens-international.org/ci2012/http:/www.citizens-international.org/ci2012/wp-content/uploads/2012/12/Development-Dictionary.pdf. Accessed 24 April 2016.

Rist, G. (2010). Development as a buzzword. In A. Cornwall & D. Eade (Eds.), *Deconstructing development discourse: Buzzwords and fuzzwords* (pp. 19–28). Rugby: Practical Action Publishing in association with Oxfam GB.

Sachs, W. (2010a). Introduction. In W. Sachs (Ed.), *The development dictionary: A guide to knowledge as power* (2nd ed., pp. xv–xx). London: Zed. [Online] http://www.citizens-international.org/ci2012/http:/www.citizens-international.org/ci2012/wp-content/uploads/2012/12/Development-Dictionary.pdf. Accessed 28 April 2016.

Sachs, W. (2010b). Environment. In W. Sachs (Ed.), *The development dictionary: A guide to knowledge as power* (2nd ed., pp. 24–37). London: Zed. [Online] http://www.citizens-international.org/ci2012/http:/www.citizens-international.org/ci2012/wp-content/uploads/2012/12/Development-Dictionary.pdf. Accessed 24 April 2016.

Samaratunga, T. C. (2013). Living skyline: High-density high-rise low-income housing: An appropriate city planning solution for Colombo, Sri Lanka? *Doctor of Philosophy Thesis*, Bond University: Institute of Sustainable Development and Architecture. [Online] http://epublications.bond.edu.au/cgi/viewcontent.cgi?article=1124&context=theses. Accessed 23 April 2016.

Servaes, J., & Malikhao, P. (2008). Development communication approaches in an international perspective. In J. Servaes (Ed.), *Communication for development and social change* (pp. 158–179). New Delhi: Sage.

Servaes, J., & Oyedemi, T. (Eds.). (2016a). *The praxis of social inequality in media: A global perspective*. Lanham, MD: Lexington Books, Rowman and Littefield.

Servaes, J., & Oyedemi, T. (Eds.). (2016b). *Social inequalities, media, and communication: Theory and roots*. Lanham, MD: Lexington Books, Rowman and Littefield.

Scoones, I. (2010). Sustainability. In A. Cornwall, & D. Eade (Eds.), *Deconstructing development discourse: Buzzwords and fuzzwords* (pp. 153–162). Rugby: Practical Action Publishing in association with Oxfam GB. [Online] http://www.guystanding.com/files/documents/Deconstructing-development-buzzwords.pdf. Accessed 25 April 2016.

Standing, G. (2010). Social protection. In A. Cornwall & D. Eade (Eds.), *Deconstructing development discourse: Buzzwords and fuzzwords* (pp. 53–68). Rugby: Practical Action Publishing in association with Oxfam GB. [Online] http://www.guystanding.com/files/documents/Deconstructing-development-buzzwords.pdf. Accessed 25 April 2016.

Smyth, I. (2010). Talking of gender: Words and meanings in development organisations. In A. Cornwall & D. Eade (Eds.), *Deconstructing development discourse: Buzzwords and fuzzwords* (pp. 143–152). Rugby: Practical Action Publishing in association with Oxfam GB. [Online] http://www.guystanding.com/files/documents/Deconstructing-development-buzzwords.pdf. Accessed 24 April 2016.

Toye, J. (2007). Poverty reduction. *Development in practice*, 17(4–5), 505–510.

United Nations. (2014). *The road to dignity by 2030: Ending poverty, transforming all lives and protecting the planet—Synthesis report of the secretary-general on the post-2015 agenda*. [Online] http://www.un.org/disabilities/documents/reports/SG_Synthesis_Report_Road_to_Dignity_by_2030.pdf. Accessed 24 April 2016.

United Nations. (2015). *The millennium development goals report 2015*. [Online] http://www.un.org/millenniumgoals/2015_MDG_Report/pdf/MDG%202015%20rev%20(July%201).pdf. Accessed 28 April 2016.

United Nations Development Programme. (2013). *Humanity divided: Confront inequality in developing countries*. [Online] http://www.undp.org/content/dam/undp/library/Poverty%20Reduction/Inclusive%20development/Humanity%20Divided/HumanityDivided_Full-Report.pdf. Accessed 20 April 2016.

United Nations General Assembly. (2000). United Nations millennium declaration. Resolution adopted by the General Assembly, 18 September 2000, A/RES/55/2. [Online] http://www.un.org/millennium/declaration/ares552e.pdf. Accessed 24 April 2016.

United Nations General Assembly. (2001). *Road map towards the implementation of the United Nations millennium declaration*, 6 September, A/56/326. [Online] http://www.un.org/documents/ga/docs/56/a56326.pdf. Accessed 29 April 2016.

United Nations General Assembly. (2015). *Resolution adopted by the general assembly on 25 September 2015—Transforming our world: the 2030 agenda for sustainable development*, 21 October, A/RES/70/1. [Online] http://www.un.org/ga/search/view_doc.asp?symbol=A/RES/70/1&Lang=E. Accessed 29 April 2016.

United Nations Sri Lanka. (2015). *Millennium development goals country report 2014—Sri Lanka*. [Online] http://countryoffice.unfpa.org/srilanka/drive/MDG-Country-Report-2014.pdf. Accessed 24 April 2016.

Religious Environmentalism and Environmental Sustainability in Asia

Anthony Le Duc

Abstract The phase of the eight millennium development goals (MDGs) adopted by the United Nations at the beginning of 2000 has given way to the program of seventeen sustainable development goals (SDGs) ratified in 2015, to be implemented until 2030. While the number of SDGs is more than twice that of MDGs and more comprehensive in outlook, examination of the individual goals indicates that most if not all of the SDGs are either directly related to environmental sustainability or indirectly concerned with the quality of the environment. This is not surprising because nations have realized, if somewhat late, that human well-being cannot be dissociated from the quality of ecosystems. The escalating global environmental crisis threatens economic and social stability and makes the innate human desire for happiness even more difficult to attain. The issue, moreover, has grown into something that cannot be confined to a single or even a few sectors of society, or that can be adequately addressed by politicians or scientists alone. Rather, achieving environmental sustainability, which is an essential component of the SDGs program, requires an interdisciplinary, dialectical, and dialogical approach involving a diverse collection of individuals, groups, organizations, and institutions. Political will, social and economic reforms, scientific and technological know-how, and religious and personal commitment are all part of the effort to address the environmental woes of the modern era. The role and contribution of religious systems and traditions for the achievement of SDGs, particularly in Asia, is the focus of this chapter. This chapter aims to present the following: (1) stating the reasons why religion is essential to the aspirations of the SDGs in Asia and (2) exploring how the major religions in Asia can contribute to promoting environmental sustainability by providing a framework for (a) assessing the root cause of environmental destruction; (b) envisioning a religious-based approach to how human beings could relate to the natural environment; and (c) presenting how religion promotes harmonious human-nature relationship through a program of

A. Le Duc (✉)
Asian Research Center for the Study of Religion and Social Communication, Bangkok 10120, Thailand
e-mail: Anthony.Leduc@Divineword.Com.Au

J. Servaes (ed.), *Sustainable Development Goals in the Asian Context*, Communication, Culture and Change in Asia 2,
DOI 10.1007/978-981-10-2815-1_3

self-cultivation and self-transformation. This chapter hopes to demonstrate that the task of achieving the SDGs in Asia is tied with the concern of religion and progress depends greatly on the improved state of human moral and spiritual well-being that religion aims to promote.

Keyword Religion · Environmentalism · Asian cultures · Inter-religious dialogue

Abbreviations

A Aṅguttara
D Digha Nikāya
Dp Dhammapada
J Jātaka
M Majjhima Nikāya
S Saṃyutta Nikāya

1 The Relevance of Religion to the Achievement of the SDGs in Asia

In 2010, Pew Research Center's Forum on Religion & Public Life published the result of its study of the global religious landscape and declared that 84 % of the world population at that time (6.9 billion) had a religious affiliation.[1] The demographic study, which analyzed data from over 2500 censuses in more than 230 countries and territories, found that Christians made up about a third of the world population and Muslims constituted nearly a quarter (1.6 billion). Hindus and Buddhists together almost equaled the number of Muslims (1 billion and 500 million, respectively). In addition to the major religions, more than 400 million people practiced folk religions found in the various continents and regions around the world. While the study found that approximately 16 % (about 1.1 billion) declared to not have religious affiliation, meaning that they did not identify with any particular faith, many of them indicated that they held religious or spiritual beliefs such as in God or some transcendent powers.[2]

The Pew Research Center also implemented other studies specifically focused on Asia and found quite interesting results. The organization estimates that by 2050, the number of Hindus which constitutes the largest religious group in the Asia-Pacific region will expand to nearly 1.4 billion. Nonetheless, Muslims which presently rank second will eventually overtake Hindus by the middle of the century and number close to 1.5 billion. The vast majority of the Buddhists in the world will

[1]Pew Research Center's Forum on Religion & Public Life (2016a, b),
[2]Ibid.

still hail from Asia even though the number will slightly drop as time goes on. On the other hand, by mid-century, Christians are expected to grow from 287 million in 2010 to 381 million. At the same time, the number of Asians identifying themselves as belonging to a folk religion will be slightly less than that of Christians. Despite the fluctuations in the rankings, the study projects that by mid-century, the number of Asians who claim a religious affiliation will not have declined. Instead, the size of the religiously unaffiliated group is expected to decrease from 860 million to about 838 million.[3]

The Asian context, however, warrants some commentary regarding the statistic stating that over 21.2 % of the people on this continent are classified as religiously unaffiliated. No doubt among the hundreds of millions of people who belong to this category are those who live in the hugely populated countries of China, Japan, Vietnam, and Korea and practice different degrees of what in the West is known as Confucianism and Taoism (Daoism). Although the modern-day discipline of religious studies in the West usually classifies the non-theistic traditions of Confucianism and Taoism as religions, people from these cultures often do not. According to Randall L. Nadeau, people in China, home to Confucianism and Taoism, are very likely to deny that they are Confucianist or Taoist and that these are religions at all.[4] For East Asians, the concept of religion is a relatively new phenomenon, having been imported from the West. The word "religion" itself was first translated into Japanese (*shukyo*) only in the nineteenth century and then later adopted by Chinese scholars in the form of the Mandarin word *zongjiao*.[5] East Asians often identify themselves as non-religious because they understand religion in the manner transmitted to them to literally mean "institutional teaching" or "school of instruction." These words conjure up images of something sectarian and organizational.[6] Nadeau writes, "Since Confucianism is pervasive an diffused—it is the air that Chinese and Japanese breathe, as opposed to a 'church' that one joins—Chinese and Japanese do not see Confucianism as a religious entity."[7] The other well-known tradition from China, Taoism, is considered by Chinese throughout history to be complementary to Confucianism. They interpenetrate each other so much that these two religious and philosophical systems may be considered two aspects of a single religious tradition.[8] The fact that there are two distinct traditions called Confucianism and Taoism more reflects the Western classification than something inherently present in the Chinese religious and cultural milieu. Moreover, the fact that Chinese may not readily identify themselves as belonging to a religion called Confucianism or Taoism or to admit that these are religions at all

[3]Pew Research Center's Forum on Religion & Public Life (2016a, b).

[4]Nadeau (2014, 21).

[5]The Vietnamese whose culture was profoundly affected by Chinese culture subsequently "Vietnamized" the Chinese word to derive the term "*tôn giáo*."

[6]Nadeau, *Asian Religions*, 21.

[7]Ibid.

[8]Ibid., 60.

does not mean that these traditions are not important in their lives. Most Chinese carry out family-oriented rituals rooted in Confucianism and Taoism such as making regular offerings to ancestors even if they do not regard themselves as Confucian or Taoist.[9] Thus, even though modern scholarly understanding of religion is broad enough to include systems that do not have beliefs in God or gods common in the West, so that various Asian traditions also fall under the category of religion, many Asians may not interpret their beliefs and practices in these ways. The Asian cultural understanding of religion and self-identification may affect the result presented by the Pew survey. In reality, the number of Asians who would be considered "religious adherents" could be much greater than what the census results suggest. Nonetheless, even without these considerations, we see that not only religions continue to be an important part of the life of people all over the world, but also they become increasingly important in the life of people on the continent of Asia, whose population is slightly over 60 % of the entire world.[10]

Despite unceasing scientific and technological progress coupled with secularization which prompted many to predict that religion will eventually be wiped out globally, the *homo religiosus* (religious man) of Mircea Eliade still makes up the majority of the world in the first century of the third millennium. While it is uncertain whether neurobiologists will ever definitively discover a "religious gene" that causes human beings to be religious by nature, empirical and historical evidence seems to point to the fundamental and universal nature of religion. No society in the past or present—even the most technologically advanced—is without the presence of religion.[11] Religion continues to be important to humanity because human beings always strive to achieve change and transformation in all the dimensions of their life. Scientific and technological advancement reflects transformation at the social and material level. However, the *homo religiosus* does not only seek these types of transformation but aspires to what Frederick Streng calls "ultimate transformation" encompassing the personal, social, political, and the cosmic transformation that changes the very core of the human being. According to Streng, religion serves as the means to this kind of transformation. He writes:

> An ultimate transformation is a fundamental change from being caught up in the troubles of common existence (sin, ignorance) to living in such a way that one can cope at the deepest level with those troubles. That capacity for living allows one to experience the most authentic or deepest reality–the ultimate.[12]

---

[9]Ibid., 4.

[10]World Population, accessed March 24, 2015, http://www.nationsonline.org/oneworld/world_population.htm.

[11]One may argue that even in societies where conventional religion is greatly diminished, much of the internalized values essential to the community are those rooted in religion. In other instances, conventional religions are replaced by other ideologies such as nationalism that arguably have some of the same characteristics as religion, such as transcendence and ritual.

[12]Streng (1984, 2).

One might argue that the kind of human transformation facilitated by religion is precisely what is needed in the face of the environmental crisis that threatens to hinder the achievement of the SDGs in Asia. However, religion is not always perceived in this manner. Oftentimes, emphasis is placed on the conflicts that stem from religious differences and intolerance, and how religious violence is the cause for the failure of particular development projects.[13] The havoc caused by the likes of the Islamic State (IS), Hindu, and Christian fundamentalists has fueled the thinking in the general population and even among academics of various disciplines that religion is particularly violence-prone.[14] Development institutions and agencies, when choosing religious partners, prefer those that are seen as having humanistic leanings without strict creeds and codes.[15] Nonetheless, as religions come in all stripes and forms and play a major part in the lives of the vast majority of the people in the world, on the practical level, they cannot be excluded from an effort such as solving the environmental crisis and promoting environmental sustainability. Religions, as particularly reflected in Catholic social teaching, often emphasize integral development. Pope Francis, for example, states in his 2014 World Day of Peace Message that authentic development is not about "mere technical know-how bereft of ideals and unconcerned with the transcendent dimension of man."[16] In some ways, religions are well suited for contributing to promoting environmental sustainability because of a number of reasons. The historian Lynn White Jr sees the relevance of religion in addressing the crisis because what people do about their ecology depends on what they think about themselves in relation to things around them.[17] According to White, "Human ecology is deeply conditioned by beliefs about our nature and destiny—that is, by religion."[18] How we interpret our own story and our destiny and how we relate to other human beings and to nature are all informed by our religious belief.[19] Because of the fundamental role of religious beliefs in human life as "primordial, all-encompassing, and unique" worldviews, they have the ability to mobilize the human will and effort in order to achieve desired transformations.[20] E.N. Anderson asserts that "All traditional societies that have succeeded in managing resources well, over time, have done it in part through religious or ritual representation of resource management."[21] The Muslim scholar, Seyyed Hossein Nasr, points out that the reality of the vast

---

[13]Ogbonnaya (2014, 55).

[14]Cavanaugh (2009, 4). Although this view of religion is more prevalent in the West, globalization facilitates its transmission to Asia so that Asians may come to view religion and even themselves in a manner previously unthought of.

[15]Ogbonnaya *African Catholicism*, 55.

[16]Pope Francis (2014).

[17]White (1967: 1206).

[18]Ibid., 1205.

[19]Tucker and Grim (1997, xvi).

[20]Ibid., xi–xii.

[21]Ibid., xviii.

majority of the peoples of the world still living within a religiously bound universe means that religious ethics remain the most practical vehicle for solving the environmental crisis. Nasr writes:

> The fact remains that the vast majority of people in the world do not accept any ethics which does not have a religious foundation. This means in practical terms that if a religious figure, let us say, a *mulla* or a *brahmin* in India or Pakistan, goes to a village and tells the villagers that from the point of view of the *Sharī'ah* (Islamic law) or the Law of Manu (Hindu law) they are forbidden to cut this tree, many people would accept. But if some graduate from the University of Delhi or Karachi, who is a government official, comes and says, for rational reasons, philosophical and scientific reasons, that it is better not to cut this tree, few would heed his advice."[22]

Martin Palmer and Victoria Finlay, in a book published by the World Bank, reiterate that intellectual awareness and scientific know-how are not enough to solve the problem. In the decades following a major gathering of representatives of governments, scientific and social institutions, and major nongovernmental organizations (NGOs) in Stockholm in 1972, much has been done to publicize the truth of the crisis. However, the problem remains and the world is not yet heading out of the dangers.[23] The authors recount that in this first United Nations meeting, scientists made powerful presentations about the consequences of destruction of rain forests by countries who were selling their resources out of poverty and opportunism. The presentations, instead of contributing to assuaging the problem, actually gave ideas to politicians and business people in a number of countries about ways to make money previously unknown to them. After this particular event, the world actually witnessed a jump in forest depletion.[24] On the other hand, religious involvement has been seen to be effective in promoting environmental agenda throughout the world. In Tanzania, for example, fishermen on an island off the country's coast changed their fishing methods to a more sustainable habit after they were instructed by their imam that the method they were presently using was destructive to the environment and went against the teaching of the Qur'an. The Muslim religious leader was able to do what government officials and international groups for years tried to accomplish without success.[25] The case of the fishermen in Tanzania and many other cases of effective religious intervention in addressing the issues of justice, peace, and environmental sustainability demonstrate that the role of religion cannot be excluded from the discourse on the analysis on globalization and sustainable human and environmental development. Max Stackhouse opines that "The neglect of religion as an ordering, uniting and dividing factor in a number of influential interpretations of globalization is a major cause of misunderstanding and a studied blindness regarding what is going on in the world."[26] As most of the

[22]Nasr (2007, 31).

[23]Palmer and Finlay (2003, xiv).

[24]Ibid., xvi.

[25]Barclay (2007)

[26]Stackhouse (2007, 57).

major world religions have their origins in Asia, and religion continues to play a vital role in the life of Asians, excluding religious involvement from the effort of achieving the SDGs on the continent is unwise and shortsighted. Fortunately, as religious leaders have begun to take more proactive roles in involving themselves in the environmental discourse, the presence of religion in the conversation has garnered more attention than before. Even in Communist China, there is a resurgence of public interest in Buddhism, Confucianism, and Daoism and how these traditions can affect the course of national development.[27] Certainly, the role of religion is not over and against the secular disciplines; rather, religion serves as part of the dialogical and collaborative effort aimed toward devising a multidimensional and effective program of action on behalf of the environment.

2 Religious Framework for Environmental Sustainability in Asia

Religious contribution to promoting environmental sustainability in Asia means that the approach must take into account spiritual and transcendental dimensions inherent to the religious worldview. Religions do not simply address sociological, psychological, scientific, or ethical questions of the matter but attempt to delve into the deepest aspects of human reality in order to uncover areas of darkness, sin, and ignorance that lead to behavior that is both self-destructive and harmful toward others. Religions provide a framework for assessing actions in light of not only our mundane goals and purposes but in the context of our ultimate and long-term desire for authentic happiness whether in this world or in the world beyond. Thus, in the face of contemporary issues, millennia-old religious systems must be examined with creative eyes for fresh insights in order to address the situation. The challenge is to be able to use the religious teachings as an effective resource for promoting environmental sustainability without compromising the integrity of the tradition. As the scholar monk Bhikkhu Bodhi states:

> If any great religion is to acquire a new relevance it must negotiate some very delicate, very difficult balances. It must strike a happy balance between remaining faithful to the seminal insights of its Founder and ancient masters and acquiring the skill and flexibility to formulate these insights in ways that directly link up with the pressing existential demands of old-age. It is only too easy to veer towards one of these extremes at the expense of the other: either to adhere tenaciously to ancient formulas at the expense of present relevance, or to bend fundamental principles so freely that one drains them of their deep spiritual vitality. Above all, I think any religion today must bear in mind an important lesson impressed on us so painfully by past history: the task of religion is to liberate, not to enslave. Its purpose should be to enable its adherents to move towards the realization of the Ultimate Good and to bring the power of this realization to bear upon life in the world.[28]

---

[27] Sawyer (2015): Kindle edition.

[28] Bhikkhu (1994).

While each religion has its own framework appropriate to its philosophical, spiritual, and, depending on the tradition, theological outlook, in this part of the essay, I would like to propose three functions of religion that can uniquely contribute to the discourse. These include the three points already laid out in the beginning paragraphs of this chapter. This section is not meant to provide an in-depth or a comparative examination of the approach of each religious tradition, rather to illustrate how different religious traditions in Asia can respond to the proposed religious-based framework. The hope is that this cursory investigation demonstrates that the inclusion of religion into the development discourse is not only practical but also foundationally crucial to the present reality.

2.1 *Assessment of the Root Cause of the Environmental Crisis*

Among the various approaches to diagnosing the environmental crisis, one that is proposed by many environmental ethicists points to the underlying problem as the existence of a militant or strong anthropocentrism leading to a conflict between human beings and nature, ultimately resulting in nature's destruction and demise. On the surface, anthropocentrism sounds harmless enough as it literally means "human-centredness." Eugene Hargrove advocates this particular take on anthropocentrism. Epistemologically, anthropocentrism is unavoidable because the world can only be perceived through the human locatedness.[29] Few scholars would argue to the contrary because we cannot perceive any other way. We may try to imagine what it is like to view the world through the eyes of a chimpanzee or a bird as a stimulating intellectual or spiritual exercise, but ultimately, the only reference that we can be confident of is our own; but even with that, there are already plenty of disagreements because points of view among us often fail to coincide. However, environmental ethicists do not take issue with epistemological anthropocentrism so much as ontological anthropocentrism in which human beings are seen as at the center of the universe or at the zenith of all creation. This attitude, environmentalists charge, dangerously leads to normative anthropocentrism in which human beings claim intrinsic value for ourselves, while the rest of nature only has instrumental value. Despite its various nuances, the popular understanding of anthropocentrism, as Katie McShane states, is "the view that the nonhuman world has value only because, and insofar as, it directly or indirectly serves human interests."[30] Thus, when the interest of human beings conflicts with that of non-human entities, priority is given to the former at the cost of the latter. This charge was made by Richard Routley in 1973 in his "last man" thought experiment in which the last surviving human being in the world would not be judged as committing any ethical violations

[29]Hargrove (2003, 175).

[30]McShane (2007: 170).

if he went about destroying all other living species in order to protect his own survival.[31]

The environmental crisis from this particular perspective stems from the fact that human beings perceive ourselves as ontologically superior to nature and act based upon this understanding. This attitude arises out of a mind-set that separates human beings and nature into an unequal duality and confers on human beings the right to dominate nature and make it subservient to human needs.[32] When we turn to religion for an explanation, the analysis differs remarkably from the philosophical approach. Religious traditions tend not to interpret the environmental crisis through a philosophical understanding of the position of human beings vis-à-vis other entities in the world, but through an examination of the internal human psychological and spiritual reality and its impact on external social problems and challenges. Buddhism, for example, divides the human situation into two states: wholesome (*kusala*) and unwholesome (*akusala*).[33] The root causes of these unwholesome states are greed *(rāga)*, hatred *(dosa)*, and delusion (*moha),* while the root causes of the wholesome states are non-greed, non-hatred, and non-delusion. All animate life is impelled by these universal forces on an individual as well as collective basis. They are the motive forces behind our thoughts, words, and deeds. Introspection tells us that the unwholesome roots also known as the poisons lead to actions that result in suffering for ourselves and others in a way that contradicts with our inner desire for happiness.[34]

In basic sense, greed is that mental state in which one is constantly preoccupied with a feeling of need and want because he feels there is a lack in his life, and since his appetite is insatiable, even when he obtains what he has desired, he continues to feel the desire for lasting satisfaction. Hatred, which in the Buddhist sense includes other negative emotions such as disappointment, despair, anxiety, and dejection, also has internal origins representing dissatisfaction toward oneself and others. Finally, delusion can be seen in the form of ignorance (*avijjā*) that leads to confusion and lack of directions. It can also lead to false views that result in ideological dogmatism and fanaticism. These three unwholesome roots manifest themselves in various degrees from mild to extreme. For example, greed may be expressed in a simple wish or in something more serious such as craving and self-indulgence. Similarly, hatred can take the form of mere dislike to something much more serious such as vengefulness and wrath. Delusion can range anywhere from dullness to conceit and ideological dogmatism.[35] These three roots are not independent of one another, but are intertwined with one another and may serve as the force to impel one another. According to Buddhism, these three poisons are found in individual mental states. However, the negative consequences are not simply confined to the

---

[31]Minteer (2008, 60).

[32]Servaes (2014)

[33]D.III.275.

[34]Thera (2008, 4).

[35]Ibid., 5.

individual, but play out on the collective level as well. A person's sense of hatred for another could lead to tribal violence and international wars. A company executive's greed could impel him to carry out business practices that promote overconsumption and materialism among the general population. This is not so difficult to see with the bombardment of advertisements every time we turn on the television, go online, or drive down the street. This situation is not limited to any particular socio-religio-cultural context. Whether in Western capitalist New York or Eastern Buddhist Bangkok, the consumer culture is strong and is the engine that makes the world goes round.

Personal and social problems in the Buddhist perspective, therefore, all have their root causes in ignorance stemming from greed, hatred, and delusion. The root cause of the environmental crisis, then, must also be considered in this Buddhist approach that diagnoses the problem from a human ethical and spiritual outlook. Pragati Sahni contends that:

> As long as the mind is influenced by the three unwholesome principles of *rāga, dosa* and *moha* or greed, hatred and delusion the human race will be stricken by environmental and other forms of exploitation, as well as selfish actions, greedy consumer cultures, dissatisfaction and other attitudes that can be looked upon as vices.[36]

Likewise, the late Thai monk Buddhadasa would remark that climate change and other imbalances in nature being experienced at this time is a result of an internal human moral degeneration that affects the external dimension of the world.[37] The breakdown in human-nature relationship is reflected in the actions and activities motivated by the three poisons that promote one-sided interests without due consideration for the well-being of others, whether it is fellow human beings or the natural environment. Philip Cafaro identifies three ways that greed brings about detrimental effects to the environment.[38] First, environmental standards are breeched when businesses have greed as their driving motivation. In order to maximize profit, businesses can easily refuse to spend money on methods and instruments to safely eliminate chemicals and wastes that are produced by their factories so as not to pollute the lakes, rivers, and air that serve the needs of human beings, animals, and plants. Second, greed can undermine the democratic process. This is most clearly seen when government leaders, in order to reap personal gains, engage in acts of corruption such as instituting laws and policies that grant privileges to entities that are in the business of making money through environmentally destructive means. Third, greed is the engine that drives overconsumption. Although overconsumption is by no means a modern phenomenon having been observed since the prehistoric era, its intensification and acceleration are notable in

---

[36]Sahni (2007, 165).

[37]Buddhadasa Bhikkhu's ideas come from a number of works (Bhikkhu 2003, 2012) that have been compiled and translated by Grant A. Olson. Olson gives the title of his translation "A Notion of Buddhist Ecology." In addition to the negative effect on nature, Buddhadasa Bhikkhu asserts that internal degeneration hinders spiritual progress.

[38]Cafaro (2005, 148–149).

the twentieth century due to multiple factors including scientific advancements alongside drastically changed social and spiritual values.[39] Due to the desire for many things, people try obtain possessions that they do not really need. Companies try to maximize their profit by selling as much of a product as possible. In order to make even more profit, these same companies have to continually put out new products and they will try to convince consumers that they must have these, even though what they have been using previously is perfectly good. At the same time, competing companies also release similar products, and again consumers are told that these are better than the other ones, either in quality or in price. Overconsumption harms nature severely because in order to produce all the things that supposedly satisfy human needs, an exorbitant quantity of non-renewable natural resources must be used.

The environment can also be harmed equally by hatred that human individuals and groups display when they institute aggressive policies that aim to protect selfish economic interests. A country or organization may employ imperialistic or oppressive tactics in order to acquire control or monopoly of natural resources to which they will exploit for economic gains. As a result, entire oil fields can be depleted and entire forests can be laid barren so that company executives and government officials can line their pockets with vast sums of money. While hatred in this militant form is obvious, there are also more subtle forms of hatred, which may not readily be perceived as such. An example of a subtle form of hatred is apathy. Even though when people are told of the immanent dangers to the environment which witnesses the rapid loss of species, the depletion of forests, and the pollution of rivers and the air, they display superficial concern but do little to change their own behavior which contributes to this destruction in the first place. In a sense, apathy could be considered a passive form of hatred that collectively contributes to environmental destruction no less than the militant expressions of hatred. In addition to apathy, one can point to negligence or simply a lack of concern as manifestations of hatred, because these attitudes also express a negative disposition toward the other. While militancy may be limited to notorious individuals, groups, organizations, or governments, apathy, negligence, and lack of concern are prevalent in the great majority of the people. This makes all people susceptible to blame when it comes to the root cause of the environmental crisis.

Delusion in the Buddhist framework is certainly a driving force behind the environmental crisis since it is the foundation for the other two unwholesome states. Sometimes referred to as ignorance or possessing false views, this is a condition where people become attached to material things, thinking that they will bring about lasting happiness for them, not realizing that all things are impermanent. Ignorance is also seen in how people with expensive possessions such as cars, homes, and mobile phones are accorded higher social status. This phenomenon, then, becomes the model for how individuals in society behave and serves as the goal that all should attain to. Thus, armed with this delusion, we keep on hoarding

[39]Meinhold (2011, 1186).

and seeking without ever attaining the satisfaction that we long for, and the search goes on. Delusion or ignorance is also played out on a social level when it is believed that in real development, economic growth is the measure of national good, that high levels of production and consumption signify higher well-being, and that the importance of unceasing GDP growth trumps sustainability.[40] Delusion in the form of having false views can also be manifested in other ways when it comes to the environmental crisis. For example, individuals and groups do not have adequate knowledge of the problem or misunderstand the issues due to absorbing one-sided information from governmental agencies or interest groups. Ignorance can be a result of denial of the magnitude of the problem based on shortsighted empirical experiences that one has without considering the larger scope of what is going on. Just because one experiences a colder-than-normal winter in a particular year, it does not mean that one has adequate evidence to reject the claim that global warming is in fact taking place. Delusion can result from ideological notions supporting the stance that human beings can exercise absolute dominion over nature according to some sort of divine ordination. Likewise, delusion can be the thinking that problems will eventually be fixed if scientists can think of clever ways to solve issues. While science does indeed play an important role in rectifying the situation, scientific solutions are only part of the overall program of action. Ultimately, it depends on everyone's commitment to change their ways and habits for the better.

The religious framework for assessing the environmental crisis as demonstrated through the approach of Buddhism, therefore, looks into the deeper reality of human moral and spiritual degeneration. The state of ignorance emphasized in traditions such as Buddhism and Hinduism finds its equivalence in the state of sin in Christianity and Islam. This state of ignorance and sin manifests itself in our relational lives with others, including the natural environment. The Ecumenical Patriarch Bartholomew stated:

> For human beings… to destroy the biological diversity of God's creation; for human beings to degrade the integrity of the earth by causing changes in its climate, by stripping the earth of its natural forests or destroying its wetlands; for human beings to contaminate the earth's waters, its land, its air, and its life—these are sins.[41]

The environmental sins which Bartholomew mentioned reflect part of what Pope John Paul II continually emphasized throughout his papacy—the culture of death. For Pope John Paul II, the culture of death includes not only the lack of respect for human life in all its stages but also the lack of respect for nature. In the encyclical *Evangelium Vitae*, he writes:

> Nature itself, from being "mater" (mother), is now reduced to being "matter", and is subjected to every kind of manipulation. This is the direction in which a certain technical and scientific way of thinking, prevalent in present-day culture, appears to be leading when

[40]Ives (2013: 546).

[41]Quoted in Pope Francis (2015, 8).

> it rejects the very idea that there is a truth of creation which must be acknowledged, or a plan of God for life which must be respected.[42]

Catholic social teaching as presented by Pope John Paul II asserts that those factors that contribute to the lack of peace are not just regional conflicts, abortion, poverty, and the like but also the "*lack of due respect* for nature, by the plundering of natural resources."[43] Similar to the other social problems, the ecological crisis, says the Pope, is a moral issue reflecting a disharmonious relationship between humanity and God. "If man is not at peace with God, then earth itself cannot be at peace."[44] Addressing the environmental crisis and issues of sustainability in the Christian framework, as in other religious systems, therefore, must first and foremost begin with realizing human spiritual failures before devising ways to rectify this situation such as implementing ecological education, taking international action, and instituting structural reforms. Only when human spiritual issues have been addressed, do the means provided by the disciplines of science, politics, economics, etc., can be wisely employed to achieve the desired results for the well-being of humanity and the environment.

2.2 *Paradigm of Human-Nature Relationship in Religious Approach*

In the previous section, normative anthropocentrism was blamed for the modern-day environmental crisis. In this anthropocentric worldview, human beings see themselves as the loci of value and perception, while all other things, including other life forms, only have value relative to how they serve the interest of human beings.[45] On the individual level, anthropocentrism is comparable with an ego-centric nature, which makes one self-centered, narrow-minded, and inequitable.[46] When human beings act strongly anthropocentric, their actions can become the cause for ill-treatment of animals and wanton destruction of nature to satisfy their own needs. Even when human beings attempt to control their decision-making process by carefully examining their felt and considered preferences taking into account a worldview derived from sound aesthetic and moral ideals, and sound scientific theories, as well as a metaphysical framework that interprets these theories,[47] there is no guarantee that when push comes to shove, nature will not lose out to the interest of human beings. A natural and seemingly reasonable reaction to the potentially destructive anthropocentrism is a paradigm that helps humans to see

[42]Pope John Paul II (2016)

[43]Pope John Paul II (1990).

[44]Ibid.

[45]Callicott (1984: 299).

[46]Sahni (2007, 80).

[47]Norton (1984: 134).

themselves not as being above, apart, or removed from nature, rather as one of the entities in nature, existing within nature in an egalitarian interrelatedness. In this worldview, nature itself has intrinsic value independent of human beings' valuation and regardless of whether it can serve their needs and interests. Deep ecology, and ecophilosophy advanced by Arne Naess, represents the holistic ecocentric perspective. Based on this ecophilosophy, human beings are encouraged to attain self-realization, a state in which they understand that they are not autonomous individuals, but fully aware that they are self-in-Self, the capital Self being nature, and the human person is but a node in the intricate web of nature.[48] The ethical imperative is that human beings must accord respect to nature and must not interfere in the workings of nature or upset its ecological equilibrium. Not only is Naess' ecophilosophy an attempt to break away from anthropocentric tendencies, but other philosophies that lean toward naturalism also set out to do the same.

These two positions present a conflict of interest because the ecocentric worldview attempts to take human beings out of the position of superiority and dominance over nature which they have been well accustomed to with continual advancement in human technological capability. Nonetheless, human beings are not always willing to simply be a small branch on the gigantic tree of existence. The result is a dichotomy between anthropocentrism and ecocentrism. The former places human beings in direct opposition to the natural world, while the latter reduces human beings to simply one among equals. Needless to say, not all adherents to religion accept the latter approach. The question worth considering is whether rejecting ecocentrism necessarily means default acceptance of extreme anthropocentrism, which condones oppression and destruction. Religious perspectives assert that there need not be a dichotomy between anthropocentric egotism and radical ecocentric egalitarianism. Scholars from the Abrahamic traditions comprising of Judaism, Christianity, and Islam believe the resources available to these traditions make it possible to envision a different kind of human-nature relationship paradigm. It must be noted, as Seyyed Hossein Nasr has done, that if one were to choose to speak of the Judeo-Christian tradition instead of considering each tradition independently, then one must also include Islam since these three make up the Abrahamic family.[49] The mentioned dichotomy can be averted by considering human-nature relationship not in isolation but also in view of their relationship with God. The paradigm that replaces the two-term metaphysics (human being vs. nature) with a three-term one comprising of human beings, nature, and God is neither anthropocentric nor ecocentric, but theocentric. In this three-term metaphysics, it is not a matter of man subject nature object or vice versa, but it is God over human being and God over nature, with the two objects of human being and nature in a mutual relationship subjected to God as the creator.[50]

---

[48]Choi (2009: 168).

[49]Nasr (1992, 96).

[50]Hoffman and Sandelands (2005: 150).

The basis for this tree-term metaphysics is derived from the Book of Genesis (canonical to both Judaism and Christianity), affirming that it was God who created both human being and nature and pronounced them to be good. The first creation myth relates that God created everything in the physical universe, including human beings, in the span of six days. In this theocentric paradigm, human beings and nature are no longer placed in positions of one over the other, but are both placed under mutual relationship enjoined together by the same creator. Like the other two Abrahamic traditions, Islam considers only the power of Allah to be absolute because he is the sole creator of the universe.

> God is the One Who raised the heavens without a pillar as you can see. Then He established his control over the realm and made the sun and moon subservient to Him. Each of them will remain in motion for an appointed time. He regulates all affairs and explains the evidence (of His existence so that perhaps you will be certain of your meeting with your Lord. (Qur'an 12:2)

This kind of understanding allows for perception of nature no longer as something to be dominated over, to be investigated and exploited, and to be made to serve the interest of human beings at any cost. Rather, the relationship between human being and nature becomes one of sibling belonging to the same father. The appreciation of this fraternal relationship could be seen in the way St. Francis of Asssisi related to the natural entities around him. In his *Canticle of the Creatures*, he reworked the symbolic images from Psalm 148 and Daniel 3:57–88 to paint a cosmological picture in which aspects of nature—sun, moon, water, air, and so forth—are joined with human beings in intimate brotherly and sisterly relationships all in praise of God. For Francis, before God, human beings and nature were not antagonists but members in a harmonious community of God's creatures. In *The Life of St. Francis*, Bonaventure wrote:

> When he considered the primordial source of all things, he was filled with even more abundant piety, calling creatures, no matter how small, by the name of brother or sister, because he knew they had the same source as himself.[51]

Francis and the Franciscan tradition has always held creation to be of moral, theological, and religious significance because not only creation reflects God, communicates God, and can help human beings to understand God, but it can also praise God independent of human beings.[52] The praising of God by his creatures is not particular to the Judeo-Christian tradition. The Qur'an also depicts God's creatures as not just inanimate objects but filled with spirit and purpose.[53] Verse 17:44 states:

> The seven heavens, the earth, and whatever is between them all glorify Him. There is nothing that does not glorify Him and always praise him, but you do not understand their praise and glorification. He is All-forbearing and All-forgiving.

---

[51]Bonaventure (1978, 250).

[52]Warner (2001: 154).

[53]Jusoff and Samah (2011: 46).

The goodness in nature which finds affirmation in the holy books of the Abrahamic traditions is also developed in the Catholic patristic tradition. Ecotheologian Jame Schaeffer has done a thorough examination of writings by patristic and medieval theologians and found abundant materials that affirm the goodness of God's creation.[54] According to Schaeffer, Christian theologians taught that all the animate and inanimate things in the universe which God created were deemed to have goodness and value. For example, in his *Nature of the Good*, Augustine of Hippo recognized "generic good things to be found in all that God has created, whether spirit or body." As for Chrysostom, reflecting on Genesis 1, he argued that since God has already deemed each type of creature to be good, no one could conclude otherwise. Thomas Aquinas likewise argued for the goodness of creatures because they had God as their ultimate source of existence.

The adoption of this perception of nature requires a cognitive shift that replaces the old paradigms of "man and nature" or "man in nature" to "man and nature in God."[55] Human engagement with nature ceases to be antagonistic and instead carried out as act of collaboration with God on behalf of nature with which human destiny is intertwined. The ethic of stewardship has its starting point in this paradigm shift. The canonical support for stewardship ethic comes from examining the two creation stories in Genesis.[56] In the first story, human is depicted as being created in God's image and allowed "have dominion over the fish of the sea and over the birds of the heavens and over the livestock and over all the earth and over every creeping thing that creeps on the earth" (ESV Genesis 1:26–27). However, in the second creation story (Genesis 2:7), God created human being by molding the dust of the earth. The Lord God formed a man out of the clay of the ground and blew into his nostrils the breath of life, and so man became a living being. Having created human being from dust, God then gave the command to "cultivate" and "care" for the land (Gen 2:15), the same land from which human being was molded.

From these two creation stories, one in which human being is depicted as created in the image of God, and the other in which human being originated from the ground, Philip Hufner developed the notion of "the created co-creator."[57] The term "the created" points to the position of human being as a creation of God, who also created all of nature. Being "the created" and molded from dust, human being is interconnected with nature. In fact, in Genesis, human beings are reminded that they will return to the ground from which they arose. "By the sweat of your face you shall eat bread, till you return to the ground, for out of it you were taken; for you are dust, and to dust you shall return" (ESV Genesis 3:19).

On the other hand, human beings, being made *imago dei* is also the "co-creator," not in the sense that they are in a position to dominate nature and exploit it however they please, but that they are charged with the responsibility to care for the

[54]Schaefer (2005: 786–803).

[55]Hoffman and Sandelands (2005, 153).

[56]Huyn, "Interreligious Dialogue," 182.

[57]Ibid, 183.

environment on behalf of God. Biblical scholars assert the terms "image" (*selem*) and "likeness" (*demut*) of God in the creation story refer to the role of human beings as reminders of God's presence in the world so as to command respect for God's authority. As mere symbols of God's presence, any notion of subjugation (*kabash*) or dominion over (*radah*) the earth in Gensis 1:28 can only be interpreted as relative rather than absolute power. The mandate of dominion over nature, therefore, is not a license to exercise unrestricted domination over creation. Rather, human beings play a role more like stewards in the household of God. Douglas Hall defines the steward as the "one who has been given the responsibility for the management and service of something belonging to another, and his office presupposes a particular kind of trust on the part of the owner or master."[58] The steward assumes his role and responsibility with an attitude of living out a vocation of representing God in the world in order to care for the earth in a relation of kinship.

In the Islamic tradition, the notion of stewardship finds its equivalence in the principle of Khalifa or trusteeship. The Khalifa is a position of vicegerency or stewardship on earth entrusted by God.[59] The establishment of this trusteeship is stated in the Qur'an, "Ah lo! Your Sustainer said to the angels: Behold, I am about to establish upon earth a Khalifa" (2:30). In verse 6:167, Allah pronounces: "It is He Who appointed you Khalifs on this earth." In the Qur'an, the word Khalifa and its plural form are referred to nine times. Of these, seven times made references to the relationship between a person, people, or human kind in general and the earth or the planet.[60]

The Islamic conception of vicegerency is applied to all the dimensions of human life in which human beings are judged on how they exercise authority over those things entrusted to their stewardship.[61] Muslim environmental scholars assert that stewardship of nature falls within this responsibility. Similar to Christian stewardship, the principle of Khalifa does not condone dictatorial attitudes toward the other or the use of natural resources that are wasteful and not in harmony with the will of the creator.[62] As Nasr declares, "In Islamic eyes, only the Absolute is absolute."[63] Rather, implementation of the responsibility requires creativity and perseverance.[64] In Islam, it is said that after Allah created the world, various creatures were asked to take on the position of trusteeship. However, none of them accepted the offer because it was seen to be too burdensome. Human beings, however, accepted Allah's request for someone to take on the onerous task.

---

[58]Cited in Butkus (2002: 20).

[59]Gada (2014: 134).

[60]Ibid.

[61]Ibid.

[62]Ibid.

[63]Nasr, "Islam," 95.

[64]Jusoff and Sahman (2011, 57).

> Lo! We offered the trust unto the heavens and the earth and the hills, but they shrank from bearing it and were afraid of it. And man assumed it. Lo! he hath proved a tyrant and a fool (33:72).

The theocentric paradigm of the Abrahamic traditions, therefore, contributes to the formulation of a practical environmental ethic by forswearing the dichotomy between anthropocentrism and ecocentrism. It is no longer a question of whether human is over nature or nature is over human, but a question of how the integrity of nature can be preserved through the effort of humanity to carry out the mandates of God. Human awareness of this mandate is the basis for the ethic of stewardship to respond to the question of ultimately who is responsible for environmental sustainability. In the end, it must be humanity that takes action on behalf of nature if environmental concerns are to be resolved. This ethical responsibility, moreover, is not grounded in human will but in the demand of God, which ensures that it will be taken seriously by adherents of the religion. As Hans Küng points out, the absolute nature of religious teachings is an essential factor in encouraging commitment to solving social problems. The authoritativeness that religion exerts on its followers is able to impel them to follow espoused "norms unconditionally, i.e. in every case and everywhere–even where they run quite contrary to [their] own interest."[65] An advantage of religion is that it is able to propose a "categorical ought" that goes beyond the finite conditions of human existence, human urgencies, and even the need for the survival of humanity.[66]

The theocentric paradigm proposed by the Abrahamic tradition represents a particularly useful way that religion frames human-nature relationship that appeals to the authority of the transcendent. Religious traditions demonstrate that a viable environmental ethic may be derived from an anthropocentric epistemology. When imbued with the proper value and standards, a human-centered epistemology does not have to descend into egocentrism and exploitation. This is the same way as one's personal worldview does not automatically lead to selfishness and egotism. Confucianism, for example, is concerned self-cultivation in order to become a fully realized person (*chun tzu*), one who is imbued with a sense of empathy/sympathy (*hsin*) for others. However, in order to develop the fullest extent of this character, the person aspiring to be a *chun tzu* must continually expand his boundary of concern until it encompasses all of humanity.[67] Huston Smith writes:

> In shifting the center of one's empathic concern from oneself to one's family one transcends selfishness. The move from family to community transcends nepotism. The move from community to nation overcomes parochialism, and the move to all humanity counters chauvinistic nationalism.[68]

---

[65] Küng (2004, 52).

[66] Ibid., 53.

[67] Smith (2009, 182).

[68] Ibid.

Scholar of Confucianism Tu Weiming asserts that the concentric circles that make up Confucianism's vision of human flourishing do not begin with the self and expand to just the world but to the entire cosmos. He writes, "We are inspired by human flourishing, but we must endeavor not to be confined by anthropocentrism, for the full meaning of humanity is anthropocosmic rather than anthropocentric."[69] For Weiming, the anthropocosmic spirit is characterized by "communication between self and community, harmony between human species and nature, and mutuality between humanity and Heaven."[70] Confucianism, therefore, like Christianity, Islam, and other traditions found in Asia can envision human flourishing in ways that promote mutuality and harmony. The SDGs in Asia are ultimately concerned with human concerns for themselves and for the world. Since religions share these concerns not only in a temporal manner but even beyond, it is imperative that the various religions dialogue with one another as well as with the other disciplines in order to move discussions on environmental sustainability from mere perception to praxis.

2.3 Religious Methods and Spiritual Motivation for Promoting Environmental Sustainability

It has been observed in many parts of the world that the religious voice continues to be able to command the attention of the people in ways that politicians and scientists cannot. While this may not be true in every case, when the religious voice is balanced, authoritative, and insightful, it can influence not only the lay public but even the experts and politicians themselves. This was evidenced by the release of Pope Francis' groundbreaking encyclical *Laudato Si* in 2015 which addressed the issue of climate change and the urgent need to protect our "common home." Although it will take years to adequately assess the encyclical's impact,[71] its immediate impression on the discourse was clearly seen. Not only did it garner worldwide interest before the Paris Climate Conference (COP21) taking place near the end of 2015, but also it proved to be a major talking point at the United Nations-sponsored event where the encyclical was referred to numerous times by world leaders in their remarks.[72] What Pope Francis has shown is that religion can make the important connections between temporal issues and essential spiritual

---

[69]Weiming (1998, 17).

[70]Ibid.

[71]There has already been quite a few panel discussions organized in academic as well as religious institutions to examine the impact that the encyclical has made on the issue of climate change.

[72]*The Tablet*, "Laudato Si' a major talking point at climate change talks in Paris," accessed March 15, 2016, http://www.thetablet.co.uk/news/2885/0/cop21-laudato-si-a-major-talking-point-at-climate-change-talks-in-paris.

values that resonate with the deepest human aspirations for personal transformation and happiness.

Ultimate transformation for people of religious faith is intimately connected to goals that are immediate to their present earthly life, but, oftentimes, includes real implications for the uncertain future after death. Religion responds to the innate human desire for eternal happiness by proposing and modeling the process of self-cultivation for adherents. Promoting environmental sustainability advocated by politicians may at times come across as a necessary strategy to get re-elected or gain power. Promoting environmental awareness by scientists may fall on deaf ears of people who have seen how experts often argue among themselves about whether there is or is not an environmental problem, global warming, or climate change. The conclusion can fall either way depending on how one looks at the situation, carries out a particular study, or interprets the data. Even Donald Trump once tweeted, "Global warming is based on faulty science and manipulated data."[73] The reality and extent of climate change may find much disagreement among experts and non-experts alike. However, when a religious leader speaks to people about the environment and makes relevant connection between human relationship to nature and their moral obligations, they are accorded a degree of credibility and respect. Even if there was no urgent environmental crisis, a religious environmentalism which promotes human-nature harmony and environmental sustainability still retains its social and spiritual value and benefit.

The religious approach to social issues, as we have seen, does not simply dwell on economic and political policies or emphasize scientific and technological fixes, but addresses the deeper dimensions of human reality. It attempts to demonstrate the intimate connection between temporal problems and moral failures that manifest themselves in particularly destructive ways on both personal and communal levels. The consequences of these failures, however, are not confined to mundane phenomena but carry their impact upon ultimate human destiny and opportunity for authentic happiness. In the field of religion, the matter of ultimate destiny of the human person is found in the area of soteriology, which occupies a central position in many world religions such as Buddhism and Christianity. Soteriology is fundamentally the study of salvation in a particular religious system. According to Steven Collins, soteriology is an "attempt to find a reflective, rationalized ordering of life, and death, as a conceptual and imaginative whole, and to prescribe some means of definitively…escaping suffering and death, and achieving a 'final' happiness."[74] In Christianity, the basis for its soteriology is the Christ event comprising of the incarnation, death, resurrection, and ascension to heaven of Jesus Christ. Through these acts, Jesus becomes the only and perfect mediator between God and humanity and serves as the path through which humanity receives salvation from the state of sin and its effects. Buddhist soteriology, on the other hand, is concerned with helping human beings attain emancipation and freedom from the experience of

[73]onforb.es/rDpyGO.

[74]Collins (1998, 22).

suffering in *samsāra*. The ultimate state of liberation is the attainment of *nibbāna*, in which one is completely liberated from the wheel of rebirth and any suffering or cravings associated with mundane existence. When one has attained *nibbāna*, all the five aggregates that make up the person are destroyed, and he is free from suffering and in a state of perfect happiness.[75] For Buddhism of the Theravada tradition, however, achievement of complete emancipation does not have to take place in a single lifetime, but over hundreds even thousands of lifetimes.[76] Nonetheless, there are other higher, albeit temporary, felicities that Buddhists aim for such as one of the various levels of heaven,[77] or even a better state of earthly life.[78]

The Hindu tradition which precedes Buddhism also upholds emancipation (*moksha*) from the wheel of rebirth (*samsāra*) as the ultimate goal for all beings. Hindu religious texts teach that there is oneness and unity of oneself with every other being in the universe. The idea that there is a unique self that one "possesses" as "my" self is an illusion. Attachment to this separate ego self causes one to be blind to the true nature of the universe, which is the unity of all things. For Hindus, liberation takes place when one's true self (*atman*) is freed from attachment to mundane and unsatisfying desires that cause it to be individuated and becomes reabsorbed into pure spirit called Brahman. When liberated, the spiritual essence (*jiva*) that underlies the ego self is united to the Brahman like a drop of water that has been finally immersed into the vast ocean. The liberated soul, no longer hampered by ignorance, now realizes that Atman is in fact Brahman, the soul of God, and that he does not have a self that exists outside of this one unified self. In its liberated form, the soul which has reincarnated over innumerable lifetimes now exists as part of a greater reality freed from difference, multiplicity, and individuation. Huston Smith describes the Hindu *moksha* as a "release from the finitude that restricts us from the limitless being, consciousness, and bliss our hearts desire."[79] As in Buddhism, Hindu spiritual awakening that leads to liberation from embodied existence is a possibility for all people. However, the task is not necessarily for everyone all at once. People do so over countless lifetimes, which see them advancing (as well as regressing) through different states of life. Though long and arduous, the faithful Hindu will do his best to strive on in persistence.

While the centrality of soteriology in Buddhism and Hinduism is obvious, as it is in the Abrahamic traditions, soteriology may not be evident in the East Asian traditions such as Confucianism. In fact, some scholars even deny that

---

[75]Kalupahana (1976, 81).

[76]Even the historical Buddha lived through numerous lifetimes before achieving enlightenment.

[77]Even in the lowest of the *devas* world, *Cātummahārājika Devas*, beings live an equivalent of 9000 human years. The beings in the highest levels of the *Devas* sphere live for tens of eons where they enjoy a myriad of sensual pleasures.

[78]Someone who is a beggar in his present life may aim to be reborn with higher social status such as a doctor or businessman.

[79]Smith (2009, 21).

Confucianism has a soteriology.[80] This denial stems from the greater emphasis by classical Chinese moral theories on social harmony than on individual redemption.[81] Nonetheless, while Confucianism does not focus on transcendence as a delivery from without, it is concerned with human destiny, moral quality, and social progress. In a utopia, the society is well ordered and individuals live in harmony with heaven and earth.[82] In order to achieve such a state of well-being, Confucianism recognizes the way of heaven as the absolute and provides for the ultimate transformation of humanity.[83] Post-Buddhist Neo-Confucianism, moreover, introduced a more individualistic concept by advocating the achievement of sagehood that manifests personal moral wisdom, perfection, and fulfillment that could be seen as similar to Buddhist enlightenment.[84] Such progress could be achieved by anyone (even women in later interpretations) through education and personal effort at self-cultivation. Thus, "salvation" does not necessarily have to be a postmortem phenomenon, but can also be found individually as well as communally in this earthly existence.[85]

Whether it is for attaining of social harmony (Confucianism), *nibbāna* (Buddhism), *moksha* (Hinduism), *Jannah* paradise (Islam), or heaven (Christianity), religious systems not only present a vision of the future, but also lay out a path for the achievement of these states of transformed life. It should be noted that the Chinese word "*dao*" which Westerners use to name the tradition of "Daoism" is simply a generic term, meaning the "way" or "path." Achieving soteriological aims may depend on the effort of the individual or of a group. It may involve different types of lifestyles and adhering to the teachings and moral codes of the particular religious system. It may require submitting oneself to divine power as well as receiving spiritual intervention from others such as supernatural beings (gods, angels, bodhisattvas, etc.). Achieving soteriological aims may also depend on ethical actions inspired by beliefs that are specific to a particular religion. It is beyond the scope of this chapter to discuss in detail the different ways that religious systems support their adherents in this endeavor. However, we can turn to Early Buddhism for an example of how one may carry out self-cultivation in order to gain a transformed state of life.

For Buddhism, the path that leads to emancipation is called the Noble Eightfold Path which combines moral virtues (*sila*) with development of concentration (*samādhi*) and wisdom or insight (*pañña*) in order to attain freedom. In the *Nidāna Sutta* of the *Saṃyutta*, the Buddha extolled the Noble Eightfold Path as the "ancient

[80] Taylor (1990, 133).
[81] Hansen (2010, 27).
[82] Ibid.
[83] Taylor, *Confucianism*, 133.
[84] Hansen (2010, 27).
[85] Ellwood and Alles (2007, 405).

road travelled by the Perfectly Enlightened Ones of the past" which leads to cessation of aging and death, volitional formations.[86] It leads to "suffering's appeasement,"[87] cessation of form, feeling, perception, consciousness, clinging,[88] and cessation of *kamma*.[89] It is the raft that takes one to "the further shore, which is safe and free from danger."[90] Of course, this further shore is none other than *nibbāna* itself. The *Sila* group consists of right speech, right action, and right living. The *Samadhi* group includes right effort, right mindfulness, and right concentration. The *Pañña* group consists of right view and right thought. These three groups represent the three stages of moral training aiming to achieve higher moral discipline, higher consciousness, and higher wisdom. The ultimate goal of the training is to attain wisdom in order to directly oppose to ignorance which serves as the cause for human suffering. However, in the process of training to achieve wisdom in order to perceive things "as they really are," the path first evolves through the training of the moral discipline, which serves as the foundation for training of concentration, which subsequently serves as the foundation for training of higher wisdom. This path of self-cultivation, says Damien Keown, "is only linear in the metaphorical sense: it does not list stages which are to be passed through and left behind so much as describe the dimensions of human good and the technique for their cultivation."[91] The end of this process of cultivation of moral and intellectual virtue is *nibbāna*, where perfection has been achieved.

One may raise the question, "How does the achievement of spiritual transformation contribute to environmental sustainability?" The answer lies in the fact that one's soteriological aims can in profound ways dictate the way one conducts his life and behaves toward the people and things around him.[92] Even in religious traditions where salvation is individualistic, the path that leads to salvation is certainly not. In the Gospel of Matthew (25:31–46), Jesus told the parable of the Day of Judgment in which those who were denied entry into the heavenly kingdom were those who did not feed the hungry, give water to the thirsty, welcome the stranger, clothe the naked, care for the sick, or visit those who are imprisoned. On the other hand, those people who did these things were welcomed into eternal life by the Judge because "whatever you did for one of the least of these brothers and sisters of mine, you did for me." Thus, spiritual goals and aspirations always involve our relational life with others. Acts that do not display care and concern for the well-being of others in many faith traditions are considered as sins deserving condemnation. The environmental crisis has also brought the environment into view as part of the moral other that must also be included in our web of relationships.

---

[86]S.II.12.

[87]S.II.15.

[88]S.III.22.

[89]S.IV.35.

[90]S.IV.35.

[91]Keown (2001, 102).

[92]Le Duc (2015, 42).

Thus, achieving spiritual progress whether through self-cultivation, through group effort, or by any means afforded by the religious tradition must ultimately form individuals who are able to enter into healthy and harmonious relationship with other people and things around them. These individuals realize ultimately that their well-being is inextricably connected to the well-being and the flourishing of others around them. As the Buddha taught in the sutras of the *Aṅguttara Nikāya*, there are four types of people in the world: those who act on behalf of oneself but not others, those who act on behalf of others but not oneself, those who act neither on behalf of oneself nor of others, and those who act both on behalf of oneself and on behalf of others. Of these four, the last type of person is considered to be "the foremost, the best, the preeminent, the supreme, and the finest of these four."[93]

Religious traditions, in the face of serious environmental concerns, must set out to form individuals and inculcate into them virtues that not only benefit their own spiritual well-being but also contribute to environmental well-being and sustainability. In the introduction to an anthology on Confucianism and the ecology, Mary Evelyn Tucker and John Berthrong write, "The ethics of self-cultivation and the nurturing of virtue in the Confucian tradition provide a broad framework for harmonizing with the natural world and completing one's role in the triad [with Heaven and Earth]."[94] Regarding Buddhism, Damien Keown writes:

> One only needs to read the *Dhammapada* to see that the Buddhist ideal of human perfection is defined in terms of the virtues exercised by an individual who treats all beings with kindness and compassion, lives honestly and righteously, controls his sensual desires, speaks the truth and lives a sober upright life, diligently fulfilling his duties, such as service to parents, to his immediate family and to those recluses and Brahmans who depend on the laity for their maintenance.… A Buddhist ecology, then, coincides with these teachings and simply calls for the orientation of traditional virtues towards a new set of problems concerned with the environment.[95]

What is true regarding Confucian and Buddhist virtues vis-à-vis the environment also finds resonance in other religious systems such as Hinduism, Islam, and Christianity. Louke Van Wensveen states clearly, ecological transformation "involves millions of Christians learning to cultivate ecological virtues as a conscious and integrated part of their Christian identity, both personal and communal."[96] What is necessary is that the original intention of the traditional virtues must be re-examined and reformulated in such a way that makes them relevant to the contemporary situation. Therefore, the exercise of the Confucian virtue of benevolence (*jen*) must not only be aimed at achieving harmony envisioned in purely social terms but also be expanded to include the all-encompassing reality of the cosmos. Similarly, the virtue of responsibility is to be upheld in view of the stewardship ethics promoted by Islam and Christianity which see human beings as

93 A 4:95.

94 Tucker and Berthrong (1998, xxxviii).

95 Keown (2007: 109–110).

96 Van Wensveen (2000, 167).

custodians rather than masters of the environment. The virtues of compassion, mercy, and loving kindness are found in both Western and Eastern religious traditions and reflect a selfless disposition toward the other. A truly compassionate person does display this virtue not only to his kinfolks but also to all—human beings, animals, plants, and even abiotic entities such as mountains, sand dunes and caves. If a person is kind and merciful toward fellow human beings and even animals, but wreaks havoc on rivers and forests, he could hardly be said to be truly compassionate. The list of virtues which contribute to environmental sustainability are as numerous as the list of virtues that help promote human development, self-transformation, and, ultimately, salvation. In other words, virtues that are deemed socially beneficial are equally ecologically relevant when the environment is intentionally made the object of one's virtuous behavior. This explains why when the exercise of certain virtues is lacking, imbalances not only are witnessed on a personal or social level, but also can be seen ecologically.[97] As Pope Francis asserts, "It is no longer enough to speak only of the integrity of ecosystems. We have to dare to speak of the integrity of human life, of the need to promote and unify all the great values."[98] Religion's task is to call attention to this unity of values as well as to persistently urge for the cultivation and exercise of these values in the daily life of the adherents.

3 Conclusion

In this chapter, it was proposed that religion is not only important but necessary to the realization of the Sustainable Development Goals in Asia. Asia is a vast continent and the home of many world religions. All indications show that the people on this continent of Asia are still religious, and will continue to be more so in the future. What is considered as "religion" is a controversial academic debate that, depending on the definition, leads to the inclusion or exclusion of certain traditions in Asia (Shinto in Japan, Confucianism in China, etc.). Nonetheless, if what is considered "religious" is seen as something opposed to be purely social, secular, or humanist, then we can argue that the overwhelming majority of Asians fall into this category. As such the exclusion of these traditions from having a role in the SDGs program means turning a blind eye on the potential impact that these traditions can exert on the people. This chapter outlined a general religious framework for promoting environmental sustainability that includes assessing the root of the environmental crisis, re-envisioning the true nature of human-nature relationship, and cultivating virtues that are beneficial to human happiness as well as environmental sustainability. This chapter was not meant to present a study of how each religion coheres to this framework since this would be far beyond the scope and length of

[97]Pope Francis, *Laudato Si*, 224.

[98]Ibid.

what a short essay can achieve. Such a task, however, is achievable if adequate effort is put in. What has been done in a cursory manner is to show that religions add a unique dimension to the discourse with their worldviews and approaches. Admittedly, in Asia, the home to many of the world religions and where all the traditions mentioned in this chapter are either playing or increasingly playing important roles in people's lives, environmental devastation has taken place as much if not more than in other places of the world. This is not reason to either place the blame for the environmental crisis on religious teachings or discount the potential of religion to contribute to addressing the situation. The environmental crisis is an opportunity and a challenge for these traditions to re-examine themselves with creative and sincere minds in order to formulate a relevant and timely response to this global problem. Certainly, the method of religion cannot replace those from science and the social disciplines; however, it can help move the discourse beyond what these disciplines can do by themselves. Thus, the achievement of the SDGs in Asia demands the collective wisdom that is available through the intervention of all human institutions and traditions.

References

Barclay, E. (2007, October 31). African fishermen find way of conservation in the Koran. *The Christian Science Monitor* October 31, 2007. Accessed 26 February 2016. http://www.csmonitor.com/2007/1031/p01s04-woaf.html

Bhikkhu, B. (1994). *A buddhist response to contemporary dilemmas of human existence*. Accessed 10 May 2012. http://www.accesstoinsight.org/lib/authors/bodhi/response.html

Bhikkhu, B. (2003). (Trans.). *The connected discourses of the buddha: A translation of the Saṃyutta Nikāya*, 2nd edn. Boston: Wisdom Publications.

Bhikkhu, B. (2012). (Trans.). *The numerical discourses of the buddha: A complete translation of the Anguttara Nikaya*. Annotated edition. Boston: Wisdom Publications.

Bonaventure. (1978). *The life of St. Francis*. Mahwah, NJ: Paulist Press.

Butkus, R. A. (2002). *The stewardship of creation* (pp. 17–23). The Centre for Christian Ethics at Baylor University.

Cafaro, P. (2005). Gluttony, arrogance, greed, and apathy: An exploration of environmental vice. In R. Sandler & P. Cafaro (Eds.), *Environmental virtue ethics* (pp. 135–158). New York: Rowman & Littlefiled Publishers.

Callicott, J. B. (1984). Non-anthropocentric value theory and environmental ethics. *American Philosophical Quarterly, 21*(4), 299–309.

Cavanaugh, W. T. (2009). *The myth of religious violence*. Oxford: Oxford University Press.

Choi, H. M. (2009). Interreligious dialogue: Toward overcoming the eco-crisis. *International Journal of Buddhist Thought & Culture, 12*, 157–192.

Collins, S. (1998). *Nirvana and other buddhist felicities*. Cambridge: Cambridge University Press.

Ellwood, R., & Alles, G. (2007). *Encyclopedia of world religion*. New York: Infobase Publishing.

Gada, M. Y. (2014). Environmental ethics in Islam: Principles and perspectives. *World Journal of Islamic History and Civilization, 4*(4), 130–138.

Hansen, C. (2010). Chinese confucianism and daoism. In C. Taliaferro, P. Draper, & P. L. Quinn (Eds.), *A companion to philosophy of religion* (pp. 23–33). Malden, MA: Wiley-Blackwell.

Hargrove, E. (2003). Weak anthropocentric intrinsic value. In A. Light & H. Rolston III (Eds.), *Environmental ethics: An antholog* (pp. 175–190). Oxford: Blackwell Publishing.

Hoffman, A. J., & Sandelands, L. E. (2005). Getting right with nature: Anthropocentrism, ecocentrism, and theocentrism. *Organization & Environment, 18*(2), 141–162.
Ives, C. (2013). Resources for buddhist environmental ethics. *Journal of Buddhist Ethics, 20*, 541–571.
Jusoff, K., & Samah, S. A. A. (2011). Environmental sustainability: What Islam propagates. *World Applied Sciences Journal, 12*, 46–53.
Kalupahana, D. J. (1976). *Buddhist philosophy: A historical analysis*. Honolulu: University of Hawaii Press.
Keown, D. (2001). *The nature of buddhist ethics*. Basingstoke: Palgrave Macmillan.
Keown, D. (2007). Buddhism and ecology: A virtue ethics approach. *Contemporary Buddhism, 8* (2), 97–112.
Küng, H. (2004). *Global responsibility: In search of a new world ethic*. Eugene, OR: Wipf & Stock Publishers.
Le Duc, A. (2015). Buddhist soteriological aims and their contribution to environmental well-being. In R. Meinhold (Ed.), *Environmental values emerging from cultures and religions of the ASEAN region* (pp. 31–53). Bangkok: P. Press.
McShane, K. (2007). Anthropocentrism vs. nonanthropocentrism: Why should we care? *Environmental Values,* 16, 169–185.
Meinhold, R. (2011). Overconsumption. In *Encyclopedia of environmental issues, revised edition*, 1186. Pasadena: Salem Press.
Minteer, B. A. (2008). Anthropocentrism. In J. B. Callicott & R. Frodeman (Eds.), *Encyclopedia of environmental ethics and philosophy* (pp. 58–62). Detroit: Macmillan Reference USA.
Nadeau, R. L. (2014). *Asian religions: A cultural perspective*. Malden, MA: Wiley Blackwell.
Nasr, S. H. (1992). Islam and the environmental crisis. In S. C. Rockefeller & J. C. Elder (Eds.), *Spirit and nature: Why the environment is a religious issue* (pp. 86–108). Boston: Beacon Press.
Nasr, S. H. (2007). Religion and the environmental crisis. In W. C. Chittick (Ed.), *The essential Seyyed Hossein Nasr* (pp. 39–42). Bloomington: World Wisdom Inc.
Norton, B. G. (1984). Environmental ethics and weak anthropocentrism. *Environmental Ethics, 6* (2), 131–148.
Thera, N. (2008). *The roots of good and evil: Buddhist texts translated from the Pāḷi*. Kandy: Buddhist Publication Society.
Ogbonnaya, J. (2014). *African catholicism and hermeneutics of culture: Essays in the light of African Synod II.* Eugene: Or: Wipf and Stock.
Palmer, M., & Finlay, V. (2003). *Faith in conservation: New approaches to religion and the environment*. World Bank: Washington D.C.
Pew Research Center's Forum on Religion & Public Life. (2016a). *The global religious landscape*. Accessed 24 February 2016. http://www.pewforum.org/2012/12/18/global-religious-landscape-exec/
Pew Research Center's Forum on Religion & Public Life. (2016b). Demographic study: Asia-Pacific. Accessed 25 February 2016. http://www.pewforum.org/2015/04/02/asia-pacific/
Pope Francis. (2014, January 1). *World day of peace message*. Accessed 26 February 2016. https://w2.vatican.va/content/francesco/en/messages/peace/documents/papa-francesco_20131208_messaggio-xlvii-giornata-mondiale-pace-2014.html
Pope Francis. (2015, March 3). *Laudato Si*. Accessed https://w2.vatican.va/content/dam/francesco/pdf/encyclicals/documents/papa-francesco_20150524_enciclica-laudato-si_en.pdf
Pope John Paul II. *Evangeliuum vitae*. Accessed 26 March (2016). http://w2.vatican.va/content/john-paul-ii/en/encyclicals/documents/hf_jp-ii_enc_25031995_evangelium-vitae.pdf
Pope John Paul II. (1990). *World day of peace message*. Accessed 25 March 2016, https://w2.vatican.va/content/john-paul-ii/en/messages/peace/documents/hf_jp-ii_mes_19891208_xxiii-world-day-for-peace.pdf
Sahni, P. (2007). *Environmental ethics in buddhism: A virtues approach*. Routledge: Taylor & Francis.

Sawyer, J. (2015). Introduction. In J. Sawyer & D. Jin (Eds.), *Ecological civilization*. Beijing: Pulitzer Center. Kindle edition.
Schaefer, J. (2005). Valuing earth intrinsically and instrumentally: A Theological framework for environmental ethics. *Theological Studies, 66*(4), 783–814.
Servaes, J. (Ed.). (2014). *Technological determinism and social change*. Lanham: Lexington Books.
Smith, H. (2009). *The world's religions* (p. 2009). New York: Publishers Group UK.
Stackhouse, M. L. (2007). *God and globalization: Volume 4 (globalization and grace)*. New York: Continuum Publishing Group.
Streng, F. J. (1984). *Understanding religious life*. Belmont, Calif: Wadsworth Pub Co.
Taylor, R. L. (1990). *The religious dimensions of confucianism*. Albany, NY: State University of New York Press.
Tucker, M. E., & Grim, J. (1997). Series foreword. In M. E. Tucker & D. R. Williams (Eds.), *Buddhism and ecology* (pp. xv–xxxiii). Cambridge: Harvard University Press.
Tucker, M. E., & Berthrong, J. (1998). "Introduction." In M. E. Tucker & J. Berthrong (Eds.), *Confucianism and ecology* (pp. xxxv–xlv). Cambridge, MA: Harvard Press.
Warner, K. (2001). Franciscan environmental ethic: Imaging creation as a community of care. *Journal of the Society of Christian Ethics, 31*(1), 143–160.
Weiming, T. (1998). Beyond the enlightenment mentality. In M. E. Tucker & J. Berthrong (Eds.), *Confucianism and ecology* (pp. 3–22). Cambridge, MA: Harvard University Press.
Van Wensveen, L. (2000). Christian ecological virtue ethics. In D. T. Hessel & R. R. Ruether (Eds.), *Christianity and ecology* (pp. 155–167). Cambridge, MA: Harvard University Press.
White Jr, L. (1967). The historical roots of our ecologic crisis. *Science, 155*(3767), 1203–1207.

Sustainable Development Goals (SDGs) in Perspective: Lessons from the American Episcopal Missions in Sagada, Northern Philippines

Danesto B. Anacio

Abstract In September 2015, world leaders adopted a set of 17 Sustainable Development Goals (SDGs) to end poverty, fight inequality and injustice, and tackle climate change by 2030. The SDGs were built upon the earlier Millennium Development Goals (MDGs) and aim to achieve universal progress without leaving anyone behind. While the SDGs provide a glorious framework for espousing development for all peoples, identifying specific nuances regarding "development" needs to be identified to contextualize the accomplishment of the SDGs, especially in the case of Asia, a multi-ethnic and highly biodiverse region. In understanding these development nuances, this chapter explores an analogy between recent and international development efforts with missionary activities initiated by American Episcopal missionaries in Sagada, northern Philippines, during the early-to-middle twentieth century. As an indigenous community, certain aspects of Sagada culture allow natives to interact harmoniously with their ecosystems in a distinct culture–environment nexus. For instance, indigenous values indicate land and water bodies as places which should be dealt with care if to avoid being a receiver of some sort of paranormal retribution. Undoubtedly, the indigenous religion, cultural, and social arrangements have enabled Sagada folk to sustain their landscapes and natural resources into the modern world amidst culture changing arrangements introduced by American Episcopal missionary works. On the other hand, unfavourable effects have been inevitable and thus need to be emphasized in order to sustain growth beyond SDGs.

Keywords Indigenous peoples · Culture change · Biodegradable wastes · Indigenous knowledge systems

---

D.B. Anacio (✉)
School of Environmental Science and Management (SESAM),
University of the Philippines Los Baños, Los Baños, Philippines
e-mail: dbanacio@yahoo.com

J. Servaes (ed.), *Sustainable Development Goals in the Asian Context*,
Communication, Culture and Change in Asia 2,
DOI 10.1007/978-981-10-2815-1_4

1 Introduction

To contribute to the current SDGs discourse, this chapter presents a reflexive approach to learn from previous development experiences (aside from the MDGs and other similar initiatives) and draw parallel scenarios which could reflect similar trends during the implementation of the 17 SDGs. Analysing the sequence of past events, for example, reveals patterns of cause and effect that would have been otherwise regarded as trivial occurrences. In the case of Sagada municipality, northern Philippines, various incidences have been in place which helped shape and enable necessary conditions for Sagada's current sociocultural and environmental landscape. Most importantly for this chapter, environmental and cultural conditions in Sagada have been greatly modified as a result of various historical events and development initiatives.

2 Setting the Context: Sagada, Philippines, and American Episcopal Missions

2.1 Sagada Socio-economic and Biophysical Profile

The municipality of Sagada, Mountain Province, Cordillera Administrative Region (CAR), is a fifth class municipality, having an average annual income of at least 15 million pesos (around 321,000 USD) and not more than 25 million pesos (around 535,000 USD) (PhP46 = 1 USD). Sagada is a landlocked municipality in northern Philippines and geographically located at 17°05′ north and 120°54′ east (Fig. 1), bordered with the municipality of Tubo, Abra, in the north; the municipality of Bontoc, Mt. Province, in the east; the municipality of Sabangan, Mt. Province, in the south; and the municipalities of Bauko and Besao, Mt. Province, in the west. It has a total land area of 9969 ha, wherein around 99.3 % is classified by national law as forest and public land, with an elevation range of 1313 metres above sea level (masl) to 2318 masl. Sagada is situated in a Type 1 climate of the four climate types in the country identified by the Philippine Atmospheric, Geophysical and Astronomical Services Administration (PAGASA), having two pronounced seasons: dry season occurs from November to April and the wet season from May to October.

As of the latest census in 2010, Sagada has a total population of 11,244 comprising 2677 households and a population density of 114 individuals per square kilometre growing at an annual rate of 0.43 % (Philippine National Statistics Office, NSO 2013). Nacagang is the least populated barangay, while Patay is the most populated (Table 1). People of Sagada commonly identify themselves as i-Sagada (/i:/-Sagada) to denote their belongingness to the place or at other times as Igolot (Igorot), a collective term ascribed to peoples of the Luzon Cordillera, or highland

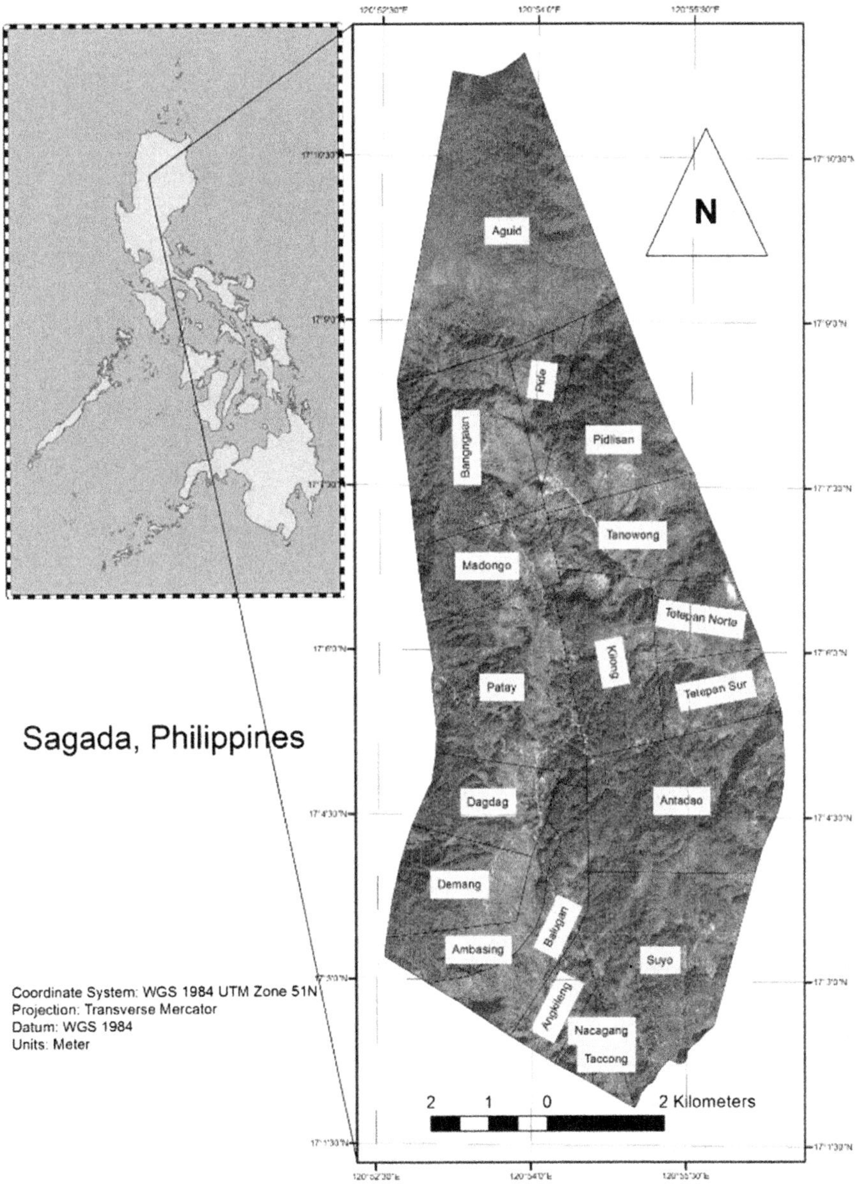

Fig. 1 The administrative boundaries of the municipality of Sagada

peoples in general. I-Sagada (as well as applied by other Cordilleran dialects) also uses the prefix *i-* (/iː/) for specifying one's ancestral *ili* (village or place of origin), for example, *i*-Pidlisan, *i*-Bugang, and *i*-Aguid.

Table 1 Land area and population in barangays (smallest administrative unit) of Sagada

Barangay name	Area (ha)	Population (2010 census)
Aguid	2717.68	596
Antadao	1405.53	360
Tanowong/Tano-ong	625.98	402
Balugan	618.60	813
Pidlisan	587.65	462
Ankileng	555.03	994
Madongo	539.83	485
Banga-an/Bangngaan	443.00	703
Tetep-an Norte	354.87	397
Ambasing	333.73	796
Patay (Poblacion)	306.95	1537
Tetep-an Sur	240.90	408
Kilong	235.36	393
Taccong	223.38	315
Suyo	217.54	395
Nacagang	184.32	254
Demang	150.81	758
Dagdag	139.91	814
Pide/Piche	87.93	362
Overall	9969.00 ha	11,244 individuals

2.2 The Entry of American Episcopal Missions in Twentieth Century Sagada

First among the missionaries in Sagada is Charles Henry Brent, appointed Missionary Bishop of the then newly formed Missionary District of the Philippine Islands, who arrived in Manila in August 1902. While exploring the mountains of the central Cordillera region from west to east, and down the Chico River in February 1903, Bishop Brent made a declaration that would later immortalize Anglicanism among the Igorots: "If I were free to do it, I would not ask for a greater privilege than to give my life for these people" (The Spirit of Missions 1903). Among those who heeded Bishop Brent's call for missions in the northern Philippines were Rev. Walter C. Clapp who settled in the nearby municipality of Bontoc and Rev. John A. Staunton Jr. who proceeded to Sagada.

Fr. Staunton Jr. arrived in Sagada in September 1904, who was later joined by his wife, Eliza Wilkie Staunton, in December. "Padre Juan" as Fr. Staunton came to be called was a graduate in mining engineering of Columbia University in 1887, a BA degree in Harvard in 1890, and ordained deacon and priest in 1892 at the General Theological Seminary of the Episcopal Church in New York. He married Eliza, a Canadian of Scotch descent and registered nurse, also in 1892 (Scott 1962a).

During the 333-year Spanish occupation of the Philippines, Sagada (and almost all of the CAR) experienced intermittent Spanish contacts and had only been recognized by the colonial government as a political unit in 1847. Spanish soldiers and lowland members of the Guardia Civil, as well as an Augustinian Priest, took residence in Sagada during the 1890s, but the Spanish occupation was short lived, especially after the Philippine Revolution took place in 1896 and the surrender of Spain to the USA in 1898 (SMDC 2007; de Villa 1999; Scott 1974).

Not all Spaniards left Sagada after 1898, however, and one notable Spanish soldier from Catalonia (a north-eastern region in Spain), Jaime P. Masferre, eventually settled in Sagada after being discharged from military service. Mr. Masferre started a plantation in Sagada and introduced persimmon (*Diospyros* spp), citrus fruits (*Citrus* spp.), coffee (*Coffea* spp.), and chayote (*Sechium edule*) in Sagada. Additionally, Masferre married a native in the person of Mercedes Cunyap Langkew (de Villa 1999; Scott 1962a).

Igorots (which include the northern Kankana-ey) have effectively warded off the establishment of concrete Spanish headquarters in CAR from the sixteenth to nineteenth century that it is surprising to note that the arrival of the American missionaries were met with least resistance. On the other hand, the Masferres have also facilitated the entry of the missionaries. The Stauntons initially stayed at the Masferre residence in Batalao, Sagada, while negotiations for establishing the Mission of St. Mary the Virgin were being done with community leaders.

Scott (1962a) notes that the promise of education is one of the considerations by the old men of the *dap-ay* (male dormitory, also a distinct structure for various socio-religious political affairs) in allowing the establishment of the Mission of St. Mary the Virgin (SMV) in Sagada. An analysis of his text, however, reveals that ongoing inter-village conflicts (since the Spanish colonial period) may have been a greater consideration for welcoming the Americans. In hopes of deterring inter-village conflict, village leaders approved a large tract of land for the Mission, not then under cultivation and outside the ritual limits of the nearest community, which was located between the villages of Sagada and their enemies in the eastern villages towards Bontoc. Fr. Staunton, together with his wife, then moved to the appropriated site and lived in an abandoned, 12-square feet shed (which used to be a goat enclosure) in the hill. For three months, they lived, taught school, conducted a dispensary, celebrated divine services, and baptized more than a hundred converts.

In establishing SMV, Fr. Staunton's technical training and expertise in various fields proved critical in an unindustrialized place where everything had to be built up from the beginning. He also participated in the manual labour required for construction and recruited his fellow-workers from all over; his artisans were Americans, Spaniards, Chinese, Japanese, other Igorots from the region, other Filipinos (most probably Ilocanos), and natives from Sagada (Project Canterbury 1923; Scott 1962a). His wife, Mrs. Staunton, proved to be Sagada's legendary healer, giving out medicine and going around the *ili* making house calls, even venturing out at night in tropical storms on horseback if needed. A large woman handicapped by the garments and undergarments of her day, she would crawl through the waist-high doorways of native Sagada houses during visits, learned to speak the local language (Kakana-ey) to inquire about her patients' relatives and in-laws, and could name all the children she attended to (Scott 1962a).

3 Sagada Indigenous Worldviews and Mission-Espoused Development

This desire by missionaries to "improve" the life of the natives must have been motivated by the simple life (as viewed by the Missionaries) in Sagada during the early 1900s. People subsisted off rice grown from irrigated terraces carved out of precipitous slopes, and sweet potatoes were grown through shifting cultivation. Few vegetables were known; meat was enjoyed in the form of chickens and pigs at the time of religious sacrifices, seasonally supported by birds, small fish and shellfish, snails, insects, mushrooms, berries, fruits, and occasionally hunted deer or wild boar. Clothing consisted of a simple loincloth for men and wrap-around skirts of bark cloth or cotton for women. Thin blankets woven of cotton thread from the lowlands provided warmth during cold weather. Almost all work was accomplished with a kind of large machete (which had only recently served also as a weapon), and some iron-shod sticks as agricultural tools. People lived in low, windowless houses with tall, grass-thatched roofs. Cooking was done on a fireplace indoors, and smoke would rise up to holes under the ridgepole, providing warmth and dryness and preventing the mildew of grain stored in the attic. The smoky house also imparted an ebon layer on all household objects which usually consist of a few clay pots, wooden bowls and utensils, bamboo containers, and woven baskets for storing a few more household items (Scott 1962b).

On the other hand, the modest life of Sagada folk and the simple forms of materials used for everyday living are accompanied with a complex set of beliefs, rituals, and social arrangements. Core to this system is the role of elders, which are generally held in high regard for practical, religious, and other day-to-day advice. An indigenous form of governance is also in place, which although having no formal organization, membership, or title, leading authorities are usually old men who have demonstrated an ability to accomplish tasks deemed important by community members. These accomplishments are usually related to ensuring general community welfare, attaining personal or communal prosperity, and conflict resolution.

After Bishop Brent's exploration of the mountains in the central Cordillera in 1903, he ordered a portable sawmill from the USA to be installed in Sagada (Scott 1962a). In 1905, a sawmill was purchased and brought to Manila to be installed in Pidlisan, near a river whose flow was strong enough to power a turbine. The sawmill and its components, the heaviest package weighing around 110 kg (250 lbs.), were shipped from Manila to Candon and carried by men for four days to Pidlisan in 1906 (Brent 1906; Project Canterbury 1923). While Brent describes the natives to be quite friendly and apparently glad that the mill will be installed in the village since, with the presence of the Americans, they would not fear their headhunting neighbours (Bontoc villages) in the east; labour required to install the sawmill was small in quantity, poor in quality, and spasmodic in action. There were enough people to do the work, but a labour predicament exists, due to what he sees as:

> …men (are) living in a primitive state their wants are few, and nothing will induce them to expend more energy than is necessary to give them food and shelter–clothing is a circumstance not worthy of notice. Money has no attraction for the savage. Having all he cares for, the world does not hold anything worth purchasing. When he succeeds in earning a few pesos he is as apt to bury it as to do else with his coin. Gaudy cloth, brass wire, a mother-of-pearl shell or some such trifle may tickle his fancy enough to drag a few strokes of spasmodic exertion from his muscular limbs… (Brent 1906).Echoing Bishop Brent's statements, Fr. Staunton reports:The Igorot in his native state has few needs and no aspirations. A rough house, which he can easily build for himself and his wife, enough rice and *camotes* (sweet potatoes) to keep him from starving, a gee string (narrow loin cloth), and in high altitudes like Sagada a thin cotton blanket. With these as the easily supplied needs the Igorot has developed no aspirations, nor ambitions, nor real character; and has been for generations the drudge of those shrewder people who could exploit his labour or passions for their own advantage. (Staunton 1915).

As mentioned previously, Brent in 1906 noted how people in Pidlisan accepted the sawmill project as a welcome development, although not due to what it was intended for (processing lumber) but in deterring head-taking expeditions from the Bontoc villages. As a result, participation through manual labour in constructing the mill is not a sound idea, perhaps out of fear in case enemies would be coming from the other side of the mountain. Additionally, participating in the labour meant they would be away from the immediate protection of the village, or in helping protect the village in case of a headhunting raid.

Another problem pertains to the manner of compensating labour, wherein the missionaries offered wage payment in the form of cash, sometimes higher than the regular wage offered for similar work. Indeed, money has no attraction for the natives during that time, not because there is not anything worthy of purchasing but rather due to differences of world views held between the Igorot and the missionary. Money simply has no practical use since essential biological needs are met not through purchases in a cash economy, but through physical work in the fields and the forests. People are embedded in a landscape wherein food, shelter, and meaning-laden objects are the essential requirements for survival and are acquired not by using money but by manual labour, traditional knowledge, and social relations. Right at the beginning, a difference in interpreting the rationale of the project is made obvious.

While it may be true that the natives have no sense of aspiration, ambition, or real character for the "development" espoused by the missionaries (Staunton 1915), the hard labour required in everyday living is naturally not something aspired for. Thus, practices which lessen the burden of working in the fields—such as the creation of strong social networks to mobilize needed manpower, or the content of traditional prayers for newly-wed couples to be blessed with numerous children (since children are seen to augment needed labour)—make up for the apparent inconveniences as viewed by the American missionaries.

The concept of formal education, however, provided a common ground for the planned projects of the SMV Mission and indigenous social life. While children could be a valuable asset to work in the field, their propensity for play makes for an undesirable work quality; thus they are usually left on their own to play with friends or explore the village landscape. Only in times of less technical labour, such as hauling, keeping wild fauna off the fields, or simple errands of relaying information or tools, they are useful, as compared to tilling land, weeding, planting, harvesting, etc. Thus, children can spend time freely unless a specific task, which usually can be achieved easily in a short span of time, is needed. Knowledge was imparted by their interaction with the village during play, while doing chores, or observing adults doing their tasks.

When the Stauntons arrived in Sagada, a new mode of imparting knowledge was introduced. It was not through the *dap-ay* in the case of proper decorum and other cultural knowledge for adolescent boys; nor the *ebgan* (dormitory for females) in preparation for married life for teenage girls; nor through everyday interaction with village mates; but knowledge was imparted in a controlled setting of information source-information recipient. Children were thus able to spend their time in the hut of the Stauntons without much objection from the parents.

With the completion of the first school building in Sagada in 1912 (The Spirit of Missions 1915), boys and girls were housed in a dormitory, which essentially mimicked the *dap-ay* and *ebgan* set-up. This is the case since traditional architecture for family houses has a limited sleeping space which only allowed parents and their babies and or toddlers (there were no bedrooms). Sleeping space for more than that is not available; thus, children who no longer require parental support for locomotion, feeding, and toilet habits are encouraged to sleep at the *dap-ay* or *ebgan*. The dormitories of the school essentially performed the same functions, and thus, no noticeable change could be seen in household dynamics. On the other hand, schooling children spent lesser time for sleeping at, or spending time in, the *dap-ay* and *ebgan*.

The educational regimen which kept children in school twelve months a year and required native children to stay in the dormitories, until, as Principal Blanche E. Masse explains, "living like an Igorot becomes for them an impossibility", further aggravated culture change (Scott 1962a). The Ilocano language was also held in high regard as a symbol of progress and civilization, which Fr. Staunton always referred to as "their own dialect". There was a standing ban on pictures of pagan ceremonies or sacrifices, and snapshots taken on the compound were required to be developed in the Mission darkroom to remove scenes considered unedifying or detrimental to Mission interests (Scott 1962a). New technologies and other non-native materials, church-related rituals, and sporting events, among others, were greatly introduced by Fr. Staunton, while attempting to suppress some aspects of native culture.

There are cases, however, wherein labour is very much needed, either in the family's present situation or in anticipation of future working conditions, and may even force parents to take back their children already housed in the dormitories. Somebang (2007) recounts one such incident during her time in the Girls' School, 1929–1940:

> The Rev. Lee L. Rose, a priest in the Mission, encouraged us to recruit more girls from our home community to join us at school. Soon, my cousin Agrifina Da-os and a friend, Tambay Kay-an… came… But after four days, Agrifina's stepfather and Tambay's brother appeared at the dormitory after breakfast… When I went to find out what news the men brought from home, I found the girls' clothes thrown aside by the door and saw the relatives, Latawan and Balangset, literally carrying the girls away… Their families kept them home for the traditional responsibilities of baby sitting and working in the fields when they grew older.

4 The Missions and Culture Change: Changing an Indigenous Ethic of Contentment

The objectives of educating the natives may have been a simple inculcation of math, English, world history, and other fields of knowledge, but a description of Fr. Staunton's deeper agenda needs critical analysis:

> Left to himself, the Igorot will never pull up; artificially pulled up, he will inevitably drop back to the plane of least resistance….not futilely to preach to him; nor to wash him, clothe him, feed him, nor to build him a better house to live in; but to get him, by any possible expedient, to feel himself the need of some of these things and to endeavor to obtain them. We are sometimes asked how we succeed in 'getting hold of' the Igorot. Our reply is that there is nothing we less wish to do; what we aim at is to encourage the Igorot 'to get hold of' us. Between these two points of view there is all the difference that there is between a well-meaning nurse holding on to a screaming child, and a screaming child clinging to its mother. Appetite, desire, aspiration, ambition in ever so small a degree, elevates the plane on which it is possible for the Igorot to live with content……in subtle ways to inoculate him with the germ of discontent, to establish in his system cravings, desires, and necessities which his savage and heathen life cannot satisfy. The second is to put the means of satisfying these desires within reach of the Igorot's own effort, to make it possible for him to live on a plane of greater satisfaction until acquirement through effort becomes a habit, living without the decencies of life a disgust, and deprivation of the luxuries (relatively speaking, of course) a discontent (Staunton 1915).

The artificiality of "pulling the Igorot up" referred to by Fr. Staunton is the formal education system, religious services, medical outreach programs, and other related missionary works. In doing so, something more is being imparted aside from Anglican values and faith: the worldview of being an Episcopal missionary, perhaps an American one. While this "Stauntonian" philosophy cannot be said to represent the universal mission and vision of the missionaries, its impact to the natives is pretty much obvious today. Natives take much pride of having not been conquered by Spaniard expeditions as opposed to other Filipinos, but at the same time, see no problem in boasting of excellent English over the usage of native language. The familiarity of American life ways and culture has led some natives to consider the relevance of the *dap-ay*'s traditional structure and its associated rituals and practices to be of minor importance. While this desire of Sagada natives (and other Igorots in general) to have an "American-ness" has indeed run its course, its ramifications for appreciating Sagada culture have mainly been on the negative.

Another manifestation of culture change pertains to housing structures and native architecture. Traditionally, native houses have no ornate furniture, and other household fittings are very simple. Cooking had been invariably done in a square ash box fireplace with no chimney, and pots were placed on top of three stones to allow firewood to burn efficiently. This allowed warmth to be trapped inside the house but soot and smoke made eyes rheumy while indoors. All of the native houses' ceilings are about three to four feet in height; thus, one needs to bend upon entering or once inside; otherwise, a visitor's head would hit the joists of the second floor, where the *agamang* (rice granary) is located. There is only one door, no windows, and ventilation is through the crevices of the horizontal, roughly hewn board walls. The floor area serves as kitchen, dining room, and bedroom. Adjoining this multi-purpose space is a wooden box for pigs to sleep in, and opposite it is a space for chicken coops which are kept during the night or on stormy days (Pacyaya 1964).

Housing structures imitated from Mission buildings, along with the increasing use of galvanized iron (GI) sheets as construction material, have allowed the elimination of the aforementioned inconveniences inherent in native houses. New house types had a chimney and/or windows that allowed smoke to escape and were bigger in size and thus allowed one to walk upright without the head-hitting joists, beams, or the ceiling. On the other hand, the new house types meant the acquisition and use of more materials for constructing needed implements, including the felling of more pine trees. Additionally, the increase in floor area of these new house types allowed children to sleep in such houses; thus, the *dap-ay* and *ebgan* were less and less utilized as dormitories. Related with changes in house architecture were changes in community social relations. House building became a much longer and more expensive proposition, although fortunately, costs are offset by the traditional practice of *ub-obbo* (Voss 1980). While the *ub-obbo* (reciprocal labour) practice has certainly remained, natives with good jobs prefer to hire labour, partly also due that their regular office job schedules make participation difficult. Still, Voss (1980) finds that in the late 1970s, many affluent people participate in such arrangements because the social pressures to do so are very strong.

Fr. Staunton's "germ of discontent, to establish in his system cravings, desires, and necessities which his savage and heathen life cannot satisfy", was coupled with the native's industriousness—resulting to a highly esteemed value and regard for education. Having a certificate and/or diploma for completing educational requirements meant escape from toiling in the fields until old age. At the same time, however, is the desire to be washed, clothed, fed, and housed in a house similar with that of the missionary. The agenda is therefore met, in the sense that the "means of satisfying these desires" are "within the Igorot's own effort", if not with a diligent and persevering exertion characteristic of daily agrarian life.

5 Developing Sagada Under the American Episcopal Missions

As an engineer, Fr. Staunton's plans were far beyond Bishop Brent's establishment of a sawmill and dispensary. His vision for Sagada included a picture of a progressive community similar to a pioneer settlement in colonial America, sturdy pine buildings with limestone foundations housing shops, stores, and schools. By 1913, Father Staunton's enterprises had become so vast that the annual report of the Mission of St. Mary the Virgin covered 21 printed pages in the Convocation Journal and was written by eight different people. From this denuded hill, started the great missionary works that would forever mark Anglicanism in Sagada. In 1915, about ten years after the installation of the sawmill in Pidlisan, one could witness from the priest's stone veranda an outstanding achievement of the American occupation in the whole Philippines:

> 80 acres (32 hectares) of activities connected by 20 miles (32 km) of telephone wire. Four stone quarries were in operation and two lime kilns; long lines of Igorots carried lumber in from the Fidelesan (Pidlisan) sawmill and a planing mill reduced it to timber, boards and shingles; electric-lighted gasoline-powered machine and carpenter shops turned out tools and furnishings. Sweet spring water was piped into the compound under sufficient pressure to make coiled fire hoses practical in many of the 20 buildings which housed the shops, stores, supplies, and considerable herd of cows, water-buffalo and horses. Vegetables were grown both by schoolboys and professional gardeners; the Mission employed a shoemaker, tailor and laundress; and schoolgirls were already producing salable lace and hand woven cloth. Photographs of the day (developed and printed locally) show American lady missionaries with pompadours pouring tea at wicker tables in rose-trellised gardens, and Father Staunton himself dictated letters to a secretary on stationery printed on his own press in an office with three telephones on his desk. Fifty apprentices were under industrial training and 150 others on the payroll, 175 school children under instruction, and the beautiful frame church where daily services were conducted listed 2000 baptisms and 600 communicants, all of whom were privileged to make purchases in the Igorot Exchange (with items such as eyeglasses, Colgate toothpaste, etc.) whose $10,000 worth of stock had been hauled in on bull carts over a trail surveyed by the Priest-in-Charge himself (Scott 1962a).

The Episcopal Church's, or more specifically its missionaries', intents to change this "primitive" character have indeed run its course, although probably, not in its original desires and specifics. Unfortunately, Sagada's progress under Fr. Staunton needed continuous input to continue. USA declared war against Germany in 1917, and funds were not easily obtained. A proposed hydroelectric plant had to be abandoned, all power equipment was stilled by the high cost of fuel, a hospital and high school building were unfinished, and Bishop Brent resigned, sending one final Episcopal communication to recommend "curtailment and retrenchment as far as possible". The government decided to limit its own construction in Bontoc, which deprived the Pidlisan sawmill of its last market, while negotiations to sell the equipment did not go well (Scott 1962a). Construction had provided steady employment for a decade, and after 8 December 1921, a small army of labourers, stonemasons, carpenters, machine operators, mechanics, carters, and printers could find no new employer for their skills. As World War II eventually involved the

Philippines, more structures in Sagada were ultimately destroyed in 1941 (Scott 1962a).

To summarize the history of American Episcopal Missions in Sagada, Scott's (1962a, b, 1974) interpretations of Spanish records and the mention of Sagada villages during the seventeenth century, as well as his acute observations during his time as a lay missionary and teacher in St. Mary's School, have all been very critical for this chapter in analysing the Missions. His manuscripts provided baseline information and gave us a glimpse of native life in Sagada prior and during their first encounters with Episcopalian missionaries.

The selection of Sagada as a bastion of American Anglicanism can be attributed to two factors. First and foremost, the rationale for the missionaries for spreading the Anglican faith in Sagada stemmed from Bishop Charles Brent's agenda to avoid proselytizing among Roman Catholics. He was more concerned with work among non-Christians: the Muslims of Mindanao, the Chinese community in Manila, and the people in the mountains of northern Luzon who had "managed to stave off Spanish military and ecclesiastical advances into their lands" (Clymer in Jones 2004). Second is the apparent welcoming of natives for the American missionaries, when Spanish missionaries have only been able to establish a short-lived mission in the late 1800s. While this would entail a lengthy discussion concerning advances in weaponry, civil and international war, intents and personalities of explorers to the highlands, probably environmental factors, and a great multitude of variables, what this chapter would like to emphasize is the cultural setting of Sagada villages during the American period. The natives were avoiding conflict as much as possible with their eastern village neighbours, and the presence of the American missionaries was seen as a deterrent for the advances of their hostile neighbours. The establishment of the Pidlisan sawmill as well as the location of the Mission quarters in central Sagada attests to this argument.

The presence of the Mission of St. Mary the Virgin in Sagada served to be the catalyst for change, and from that point on, a multitude of changes followed. Although the Episcopalian Missions were said to be critical towards American imperialist goals, and opting to build a spiritual if not a social shield from the American empire through preaching, catechesis, education, and various types of agriculture and industrial training, hoping that the natives could meet "civilization" head on and not be corrupted by it (Jones 2002, 2013), they have nonetheless transmitted their American-ness to the Sagada natives. Introduced technologies, such as the innovation of piping water which changed the old method of fetching water from earthen or clay pots from open wells or water bodies (Pacyaya 1964), lime kilns, the sawmill, and other non-native technologies, have all been experienced and appreciated by the local people.

Reinforcing conditions and the breadth of influence brought by the missionaries have affected almost every aspect of Sagada native life. Fr. Staunton's approach in proselytizing natives and the introduced educational system further aggravated the impacts of culture change. His missionary ethic established concepts of appetite, desire, aspiration, and ambition as important values to be inculcated to the natives by "letting the people do it" or letting the natives discover these values. As an example, the Mission did not encourage the natives to give up their traditional attire but ensured that they have the means of getting non-native clothing with their own effort. This "means" refer to the shops, mills, kilns, trades, gardens, and industries set up by the Mission to provide opportunity for the gratification of new needs. "Labour begets skill, self-reliance, health, character, and—with Christ—happiness" (Staunton 1915).

Old men have even come to a point where the priest is seen as a great public benefactor and a loving individual concerned with the well-being of the community, albeit his apparent detachment from Sagada culture. American visitors and government officials see Father Staunton in a similar regard, especially since he appears as a charming host and a witty conversationalist: the dignity with which he conducted his priesthood and the aura of sacred mystery which he imparted to the sanctuary of his church; his aloofness from village affairs; failure to lay hands on the pigs, chickens, or women; his stern and not easily approached reputation by his office boys; and his implemented projects which increased Sagada prosperity all enhanced his god-like status (Scott 1962a).

6 Implications of Sagada's Mission History and the SDGs

A number of general ideas can be derived from the rich narrative of the American Episcopal Missions in Sagada in relation to the recently adopted SDGs, as well as the implementation of other development initiatives. First and foremost would involve the role of communication. While the seemingly seamless entry of the missionaries could be attributed to the end of the Spanish colonization period, the presence of the Masferres, ongoing inter-village conflict, and other probable factors not mentioned and considered in this chapter, proper communication must have been established between the goals and intents of the missionaries and the natives of Sagada. The establishment of rapport and goodwill between the missionaries and the natives allowed further missionary activities to flourish.

Second point involves the attitude and culture of actors involved in the development initiative. Aggressive yet patient and charismatic yet uncompromising agents with the drive to push for the achievement of goals and objectives are indispensable for project implementation. The Stauntons would be the prime parallel for such analogy, and in the case of SDGs, project implementers, whether on a personal and organizational scale or entity, need to be committed and dedicated in pursuing their objectives. Mrs. Staunton, who took the patience and time in knowing native way of life, was complemented by Fr. Staunton's stern and authoritative approach.

It can be argued that the missionaries were able to introduce needed development to improve native life in Sagada during that period. It was unfortunate, however, that an explicit disregard for Sagada cultural practices was present, although probably not by all personnel involved with the Missions. The educational regimen which aimed to acculturate a new culture for Sagada schoolchildren, the use of a cash-based economy for the flow of goods and services in Mission activities, and various culture changing practices have all been geared for the integration of the native to "modern" life. These approaches towards development without considering cultural identity is in fact contradictory (de Leon 2011).

Although these implications are general in form, qualitative in nature, and based primarily from a single case, these are nonetheless very important and relevant for enriching development narratives which could ultimately be used for improving the SDG framework. The main argument, however, especially for this particular case, would centre on the perils of introduced and idealized models of development. While this chapter agrees that there is no universal framework to development, including the SDGs, this does not mean that development per se is not an aspiration that society should pursue. Something more needs to be done, and aiming to achieve development targets, requires a culture-based approach to contextualize development rooted within a particular society's specific values and institutions (de Leon 2011).

Fortunately for the case of Sagada, adaptations have been made by the natives since strong cultural practices are very much present. Currently, the *dap-ay* is still functioning and recognized in community affairs, although not as strong as before; *ub-ubbo* (reciprocal labour) is still practiced by a few, and a number of indigenous community rituals and celebrations are still present; even though natives wear modern clothing, speak good English, and attend mass during Sundays. Culture change, however, does not only stem from the American Episcopal missions, and it is hoped that Sagada would aim to preserve and strengthen its rich cultural heritage into the modern world while achieving targets indicated in the 17 SDGs (CDP 2015; Ford 2015; Hak et al 2016; Sachs 2012; SDSN 2015; UN 2014; UNSC 2015).

References

Brent, C. H. (1906). A missionary sawmill. *The Spirit of Missions, 71*(10), 837–839.

CDP. (2015, June). *Transitioning from the MDGs to the SDGs: Accountability for the post-2015 era*. CDP Background Paper No. 25, ST/ESA/2015/CDP/25. United Nations Committee for Development Policy.

de Leon, F. M. (2011). *In focus: Culture in development planning. Philippine National Commission for culture and the arts*. Retrieved from: http://ncca.gov.ph/about-culture-and-arts/in-focus/culture-in-development-planning/

de Villa, J. G. (1999). E. Masferre: A tribute to the Philippine Cordillera. Asiatype Inc. and Brier Projects Inc.

Ford, L. (2015). *Sustainable development goals: All you need to know. The guardian*. Retrieved from: http://www.theguardian.com/global-development/2015/jan/19/sustainable-development-goals-united-nations

Hak, T., Janouskova, S., & Moldan, B. (2016). Sustainable development goals: A need for relevant indicators. *Ecological Indicators, 60*, 565–573.

Jones, A. W. (2002). A view from the mountain: Episcopal missionary depictions of the igorot of Northern Luzon, The Philippines, 1903–1916. *Anglican and Episcopal History 71*(3).

Jones, A. W. (2004). Working out the mission theory of Bishop Charles Henry Brent. *Missiology: An International Review, 32*(4), 421–429.
Jones, A. W. (2013). Pragmatic anti-imperialists? Episcopal missionaries in the Philippines, 1933–1935. *Anglican and Episcopal History, 82*(1), 1–28.
NSO. (2013). *Philippine National Statistics Office 2010 Census of Population and Housing*. Retrieved October 2013 from: http://www.census.gov.ph/content/population-mountain-province-showed-increase-14-thousand-results-2010-census-population-and
Pacyaya, A. (1964). Acculturation and culture change in Sagada. *Silliman Journal, 10*(1&2), 14–27.
Project Canterbury. (1923). *Handbooks on the missions of the Episcopal Church. No. III.* Philippine Islands. New York: National Council of the Protestant Episcopal Church Department of Missions. Retrieved from: http://anglicanhistory.org/asia/ph/missions1923/
Sachs, J. D. (2012). From millennium development goals to sustainable development goals. *Lancet, 379*(9832), 2206–2211.
Scott, W. H. (1962a, December). "Staunton of Sagada: Christian Civilizer." *Historical Magazine of the Protestant Episcopal Church 31*(4), 305–339.
Scott, W. H. (1962b). Cordillera architecture of Northern Luzon. *Folklore Studies, 21*, 186–220.
Scott, W. H. (1974). *Discovery of the Igorots (revised edition)*. New Day Publishers. Quezon City. ISBN 971-10-0087-3.
SDSN. (2015, June 22). *Indicators and a monitoring framework for the sustainable development goals. Launching a data revolution for the SDGs*. In: A report by the Leadership Council of the Sustainable Development Solutions Network. Revised working draft.
SMDC. (2007). Sagada Municipal Development Council (SMDC) Comprehensive land use plan (CLUP). Local Government Unit of Sagada.
Somebang, E. (2007). *The Odyssey of an Igorot mission girl*: A memoir. Igorot Heritage Press
Staunton, J. A., Jr. (1915). An opti-pessimistic outlook. *The Spirit of Missions, 80*, 751–755.
The Spirit of Missions. (1903, May). The progress of the kingdom, Philippine Notes: Northern Luzon. *The Spirit of Missions, 68*(5), 299.
The Spirit of Missions. (1915, January). Lives that have helped. *The Spirit of Missions, 80*(1), 279–285.
UN. (2014). *Millennium development goals report 2014*. United Nations, New York.
UNSC. (2015). *Technical report by the Bureau of the United Nations Statistical Commission on the process of the development of an indicator framework for the goals and targets of the post-2015 development agenda (Working draft)*. Retrieved from: https://sustainabledevelopment.un.org/content/documents/6754Technical%20report%20of%20the%20UNSC%20Bureau%20%28final%29.pdf
Voss, J. (1980). *Cooperation and market penetration: Indigenous and institutional forms of cooperation in the Cordillera Central of Northern Luzon*. VRF Series 75. Tokyo: Institute of Developing Economies.

Companies' Accountability in Sustainability: A Comparative Analysis of SDGs in Five Countries

Kamala Vainy Pillai, Pavel Slutsky, Katharina Wolf, Gaelle Duthler and Inka Stever

Abstract As global environmental and humanitarian issues exacerbate, leaders and nations are striving hard to tackle these challenges at a global level collaboratively. With the setting of the Sustainable Development Goals (SDGs) and its 2030 Agenda, the role of business in economic, social and environmental development has never been more imperative. In fact, companies have connected sustainable development (SD) practices to their own business initiatives. To support this trend, the Organization for Economic Co-ordination and Development (OECD) has recently developed a set of guidelines for the private sector to initiate responsible business practices. Globally, the trend points to businesses being more accountable, responsible and putting more emphasis on sustainability. Motivations to engage in sustainable practices vary. When companies integrate sustainability into their business, the management of these initiatives needs to be integrated as well. Certain tools and guidelines exist for companies to help support the internal management of sustainability, such as the Global Reporting Initiative (GRI). However, there is little research on how companies formulate their SD goals and objectives, set priorities

K.V. Pillai (✉)
Faculty of Business and Humanities, Curtin University—Sarawak, Miri, Malaysia
e-mail: kamala.pillai@curtin.edu.my

P. Slutsky
Communication Arts Department, Chulalongkorn University, Bangkok, Thailand
e-mail: pavel.slutsky@gmail.com

K. Wolf
Curtin Business School, Curtin University, Perth, Australia
e-mail: K.Wolf@curtin.edu.au

G. Duthler · I. Stever
College of Communication and Media Sciences, Zayed University, Abu Dhabi, UAE
e-mail: Gaelle.duthler@zu.ac.ae

I. Stever
e-mail: inka.stever@zu.ac.ae

J. Servaes (ed.), *Sustainable Development Goals in the Asian Context*, Communication, Culture and Change in Asia 2,
DOI 10.1007/978-981-10-2815-1_5

and measure the impacts of these initiatives. Doing so would increase their transparency and help convey meaningful information to their stakeholders. This chapter provides a critical analysis of how companies from five different countries integrate SDGs into their business activities. Each of these countries faces different, as well as similar types of sustainability challenges. The five countries that have been included in this research project are as follows: Thailand, the United Arab Emirates, Qatar, Australia and Malaysia. The authors conclude that the private sector lacks strategy when they plan for integrating sustainability.

Keywords Measurement · Accountability · SDGs · Sustainability · Reports · Management

1 Introduction

As businesses today operate in volatile, social and environmental peripheries—fuelled by unprecedented global climate conditions, turbulent social trends and diminishing resources, the concept of corporate sustainability (CS) continues to bring new meaning and heightened expectations. The adoption of innovative business practices, such as renewable energy, green technologies, sustainable financing, corporate governance, sustainable supply chain management and fair trade practices as the 'new wave' of the corporate mantra, is indicative of the growing importance of sustainable development approaches in business.

The concept of 'sustainability' was initially introduced in 'The World Conservation Strategy' report in 1980 (IUCN et al. 1980). It was further deliberated in the Brundtland Report in 1987 at the World Commission on the Environment of the United Nations. The Organization for Economic Co-ordination and Development (OECD 2001) defines sustainable development as 'meeting the needs of the present generation without compromising the ability of future generations to meet theirs'. The concept has since metamorphosed and is today synonymous with strategies, tactics and practices that focus on ameliorating social and environmental considerations as an integral part of an organization's actions and decisions; hence, the growing focuses on triple bottom line reporting across countries, industries and sectors.

Concomitantly, the implementation of various sustainability guidelines, frameworks and standards, such as the United Nations Global Compact (UN Global Compact 2016), the International Labour Organization (ILO) Conventions on Workplace Practice, the OECD Guidelines for Multinational Enterprises, the Global Sullivan Principles, AccountAbility AA1000S, Social Accountability SA8000 and the Global Reporting Initiative (GRI), by both public and private agencies, further attests to this global trend and industry demand. Concepts such as the triple bottom line, corporate social responsibility, corporate governance and corporate citizenship explicitly reject the traditional notion that firms only exist to make profits (Svensson and Wood 2011). Once regarded as 'voluntary extra' for a business' image, the relevance of corporate sustainability (CS) reporting is growing indubitably.

In addition to voluntary sustainability commitments, government regulations and requirements by stock exchanges are increasingly imposing mandatory reporting for companies. Earlier studies (Willard 2007, Malaysian Accountant 2007) have identified ten major market forces that are driving businesses to embrace sustainability. The drivers are grouped into two dimensions: (1) mega issues; and (2) stakeholder demands, as shown in Table 1.

1.1 New Global Force

In 2015, corporate sustainability took on a profound meaning at a global scale, with the launching of the United Nations' Sustainable Development Goals (SDGs) in Paris (World Business Council for Sustainable Development WBCSD 2015, United Nations 2015). With scientists raising the red flag on global warming in the age of Anthropocene, the involvement of business in sustainability practices has never been more compelling. In fact, business participation is considered as an imperative proponent to influence both public and private sector attitude, as well as mobilize collective change in markets and societies towards sustainable practices across the planet (United Nations 2015). Current discourse on corporate social advocacy, CEO activism and corporate activism (Charterrji and Tofel 2016, Dodd 2016) phenomenon is further indicative of the inherent influence that business may have over public opinion and consumer attitude.

1.2 State of Corporate Sustainability Reporting in Asia

Incontrovertibly referred as the 'factory of the world', the rapid development of production facilities in Asia has been heavily reliant on Western markets for more than five decades. As multinationals in Asia compete to gain access to global market shares and attract investors, the pressure to ameliorate corporate sustainability performance has never been so intense. With heightened scrutiny from green institutional investors and their strict ethical investment guidelines, business in Asia

Table 1 The 10 major market forces

Mega issues	Stakeholder demands
Climate change	'Green' consumers
Pollution/health	Activists shareholders
Globalization backlash	Civil society/NGOs
The energy crunch	Governments and regulators
Erosion of trust	Financial sector

Source Malaysian Accountant (2007)

continues to face an uphill challenge. According to the Association of Sustainable and Responsible Investment in Asia (ASrIA 2015), while Asia may be home to some of the richest people on the planet, it is also home to the largest proportion of people living in poverty. The region is, without a doubt, a complex one and full of contrasts. On the one hand, Asian nations continue to push the frontier of innovation as they tackle extreme poverty issues. This includes investments in innovative green technology from low-cost solar panels for rural green energy solutions and the cheapest mobile phones on the planet to flood-resistant crops. On the other hand, the region also exhibits human rights deficiency, ranging from foreign worker exploitation, sweatshop violations and explicit gender discrimination to human trafficking issues rising at feverish levels. Ongoing media reports on non-compliance, or industry violations, essentially bring to light the pervasive corruption in this region as well as the limited resources of the developing continent's capacity to enforce compliance (Sharma 2013). As global stakeholder pressure rises, social media exposés on blatant community-related violations and biodiversity degradation continues to haunt both businesses and governments in this region. From an international trade perspective, despite a lower global economic growth forecast post-2015/post-2016 globally, Asia's influence as a major contributor to global growth cannot be underplayed (Deloitte 2015). Hence, the spotlight on Asia has never been more exigent.

In the context of corporate sustainability (CS), over the last few years, a dramatic increase in CS adoption and reporting has been noted among companies in Asia. According to KPMG (KPMG International 2013), almost three quarters (approximately 71 %) of companies based in the Asia-Pacific region publish corporate sustainability (CS) reports. This represents a 22 % increase compared to the region's performance in 2011, when less than 50 % reported on their performance.

This is significant among emerging markets in Asia, particularly as a competitive strategy to increase business growth opportunities. Emerging markets are countries that are undergoing fast growth through rapid industrialization and economic development. These markets may not be fully developed markets but have several characteristics of a developed market with strong potential of becoming 'big players' on the world stage. The International Monetary Fund (IMF) surmises emerging market economies have higher growth—two to three times faster than their developed counterparts. In fact, business reports by Bloomberg, IMF forecasts and statistics published by the World Bank note that Asian nations have dominated the list of emerging markets (Bloomberg 2013).

In the light of the growing evidence of sustainable practices and reporting by businesses in Asia, the pertinent question that arises is how the quality of corporate sustainability (CS) reporting of companies in emerging markets is compared to that of developed economies? The authors argue that there is a strong case to examine the state of CS reporting as this study aims to track the extent of sustainability performance in emerging markets in order to facilitate the diagnosis of inherent gaps for the achievability of United Nations' SDGs by 2030.

Before the objectives, research questions and the methodology of this study are expounded, and a short overview of existing literature on corporate sustainability is deliberated in the following section.

2 Review of Literature

Extensive empirical research on corporate sustainability reporting has been undertaken for over four decades. Not surprisingly, most of these studies concentrate on companies in fully developed countries, such as those in the European Union (EU), the United States of America (USA) and Australia (Higgins et al. 2015). An analysis of studies undertaken in Asia reveals that the number of studies on CS reporting in this region is substantial. Most studies on CS reporting have focused on a content analysis of annual reports, sustainability reports and corporate websites, as well as examined internal determinants (Fifka 2013; Sharma 2013). Further, transcontinental research on Asia as a region only emerged over the recent decade (Chambers et al. 2003; Newson and Deegan 2002; Welford 2004; Williams 1999). In contrast, CS studies comparing the European and American continents, examining different sustainability-related context and practices, have captured performance over more than three decades (e.g. Fifka 2013; Gray et al. 1990; Hartman et al. 2007; Holland and Foo 2003; Maignan and Ralston 2002; Smith et al. 2005). This study seeks to expand the body of knowledge on corporate sustainability to the wider Asian region by undertaking a transcontinental approach, while focusing in particular on the state of CS reporting in emerging markets.

Transcontinental research involves studies on countries in more than one continent. There are different approaches to transcontinental studies (Fifka 2013); for instance, a global approach (e.g. KPMG 2013; Kolk and Perego 2010; Kolk 2008; Tsang et al. 2009; Morhardt 2010; Welford 2004), an exclusive focus on emerging markets in different continents (e.g. Baskin 2006) or on individual continents such as Australia (e.g. Chen and Bouvain 2009; Newson and Deegan 2002). In the Asian context, examples of transcontinental studies include the work by Chapple and Moon (2005), which investigated corporate website reporting by 50 companies in seven south-eastern and Southern Asia countries. The study's findings delineated three pertinent attributes: (1) CS reporting differed substantially across countries, (2) reporting standards were largely attributed to the national business systems; and (3) the extent of internationalization as Asian companies grew into multinationals influenced corporate attitude. Findings from most studies also reveal economic development as the primal focus of business within these countries, while social and environmental dimensions are often neglected. Further studies affirm that impact on social and environmental dimensions in CS reporting has not been considered vigorously in developing and emerging markets at all, compared to highly developed countries (Sharma 2013; Fifka 2013). In addition, while emerging markets have the potential to become global players and compete with companies from fully developed nations over the coming decade, there is little evidence of a definitive inquiry on the current state of CS reporting in terms of significant differences between emerging and developed markets. Therefore, the authors argue that there is a gap in the existing literature that demands further investigation.

The objective of this study is, therefore, to conduct a comparative study of the state of corporate sustainability (CS) reporting in the wider Asian region by the largest companies in selected emerging markets, versus companies in a developed nation to determine (1) whether variations exist in CS performance; (2) the extent of variation if they exist; and (3) the variables that can help explain the existing variations.

In the context of this study, the representation of the Asian region is based on UN's 2014 country classification report (UN 2014) that geographically clusters countries into three major groups: (1) East Asia, (2) South Asia and (3) West Asia (UN 2014). Both East and South Asia are also part of the Asia-Pacific region. Next, various emerging market indices are sourced to identify the emerging markets in this region (see Table 2).

In determining the emerging economies for the purpose of this study, a review of reputed business and financial media and indexes was undertaken. The findings illuminate new insights. Firstly, there is a growing trade interest in emerging nations in Asia other than BRICS (e.g. India, China) nations. Secondly, based on rankings of emerging markets (Fortune 2015; Bloomberg 2015), Asian economies listed in the top 5 include South Korea, Malaysia and Thailand. Among the West Asian emerging nations, the United Arab Emirates (UAE) and Qatar were also ranked as the most promising frontier markets (Bloomberg 2014). Frontier markets represent countries that are more profitable to invest. Hence, for the purpose of the study and based on the evidence, four top emerging markets were identified from this region: (1) Malaysia, (2) Thailand, (3) United Arab Emirates (UAE) and Qatar. Next, in determining the developed economy for the purpose of study, the criteria of selection were (1) a developed economy in Asia; and (2) one that has a breadth of industries similar to the selected emerging economies to ensure congruence in analysis. Based on the World Bank's (2015) databank statistics, Australia was selected for this study as it, firstly, met the study's selection criteria, secondly, its geopolitical location in Asia and Pacific and furthermore, as Australia is a country and a continent in itself; it provided the authors with an opportunity to conduct a transcontinental project.

Table 2 Emerging markets in Asia 2015

East Asia	China, Indonesia, Malaysia, Philippines, South Korea, Taiwan, Thailand
South Asia	India
West Asia	Qatar, Turkey, United Arab Emirates

Source Various Emerging Market Indices (International Monetary Fund IMF, Financial Times Stock Exchange FTSE, Morgan Stanley Capital International MSCI, Standard & Poor's S&P, Emerging Markets EM Bond Index, Dow Jones, Russell Index)

The study addressed the following research questions (RQ):

RQ1: How do the top 25 companies in each of the selected countries publicly disclose their approach to formulating measurable and time-framed sustainable development goals?

RQ2: How are sustainable development initiatives reported in terms of accountability to stakeholders?

RQ3: How are sustainability development initiatives measured?

RQ4: How successful are companies at reaching their sustainable development goals?

RQ5: How do the five countries compare to one another?

In the next section, theoretical approaches guiding this study are discussed, followed by explications on the methodology adopted.

3 Discussion of Theory and Measurement of Sustainable Development

Sustainable development is a concept that can be approached from various methodological positions. Based on the premise by Garriga and Mele (2004), three main theoretical approaches can be applied. First, instrumental approaches treat sustainable development as a means to achieve business goals or to improve a company's intangible indicators, such as image or reputation. In this context, sustainable development is considered as an instrument to achieving a competitive advantage (inter alia creating greater consumer support, reducing a negative public image or attracting manpower). Second, a political approach to sustainable development concentrates on the relationship between business and government structures, which may involve the voluntary (or involuntary) compliance to sustainable development guidelines that present companies with a better position in government relations and reduce vulnerability to drastic actions by the state (Baeten 2000; Bernstein 2002; Demeritt 2006; Dovers 1996). Finally, a strategic development approach to sustainability examines the link between strategic requirements to business operations and concerns for future developments that focus on sustaining finite resources in order to provide for the future needs of the company (Porter and Kramer 2006). Each of these approaches facilitates the evaluation of CS reporting by linking the reporting of companies' sustainable development practices with their official declarations on sustainable development.

4 Materials and Methodology

This transcontinental study set out to examine CS reporting of major companies in four emerging economies and one fully developed economy in the Asian region. The basis for the analysis is the annual reports of the 25 largest companies of each country. The selection of companies was based on market capitalization, as all companies traded on the stock market are obligated to publish annual reports. This enabled the study to investigate the general trend of listed companies across various sectors towards sustainability practices. If applicable, stand-alone sustainability or corporate social responsibility (CSR) reports were included in the data collection. Out of the total sample size of 125 reports, 72 were annual reports (59.51 %), 38 were sustainability reports (31.4 %), while the remaining number comprised of corporate social responsibility or corporate responsibility (CR) reports, as well as financial reports (Fig. 1).

Industry representation: the majority of companies represented (Table 3) are from the banking (24 %), energy (11.2 %), real estate & construction (11.2 %) and telecommunication (9.6 %) sectors, due to the nature of their national economies.

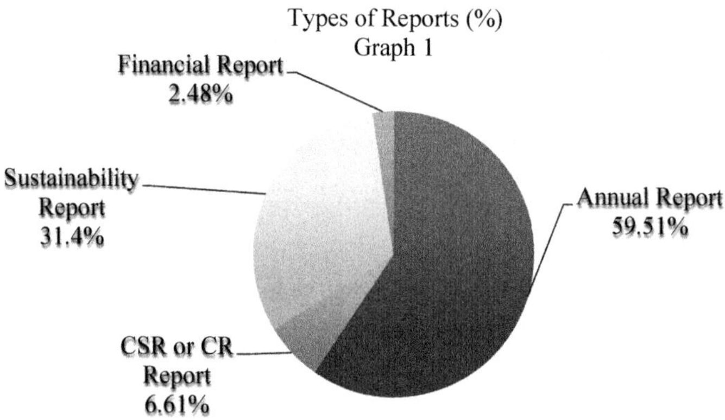

Fig. 1 Types of reports (%)

Table 3 Industry represented in sample

	F	%
Agriculture	5	4.0
Banking	31	24.8
Conglomerate	5	4.0
Energy	14	11.2
Financial services	9	7.2
Food	2	1.6
Health care	5	4.0
Hospitality	2	1.6
Industrials	12	9.6
Real estate and construction	14	11.2
Retail	3	2.4
Telecommunications	12	9.6
Transportation	7	5.6
Utilities	4	3.2
Others	5	4.0

To ensure an objective analysis of the reports across countries, a coding system was developed. The coding was done by three researchers independently. The intercoder reliability was 96.74 %. The coding categories were as follows:

1. Country
2. Name of company
3. Market Capital
4. Industry
5. Types of report: 1 = annual report, 2 = CSR report, 3 = sustainability report, 4 = financial report, 5 = other
6. Sustainability in report? 1 = yes, 2 = no
7. Sustainable goals stated: 1 = yes, 2 = no
8. Time frame included: 1 = yes, 2 = no
9. Measurable (quantitative parameters): 1 = yes, 2 = no
10. Type of SD initiative Environmental: 1 = yes. 2 = no
11. Type of SD initiative Social: 1 = yes, 2 = no
12. Type of SD initiative Economic: 1 = yes, 2 = no
13. Guidelines followed? 1 = yes, 2 = no
14. Multiyear performance track record: 1 = yes, 2 = no
15. Objectives measured (objectives vs. results): 1 = yes, 2 = no
16. Independent assessment: 1 = yes, 2 = no

5 Results

RQ1: How do companies publicly disclose their approach to formulating measurable and time-framed SD goals?

Results in Fig. 2 indicate that the majority of companies engage in and report on sustainability. However, a minority of the remaining major companies do not report or engage, indicating a lack of concern for implementing or improving the quality of their triple bottom line (TBL) reporting.

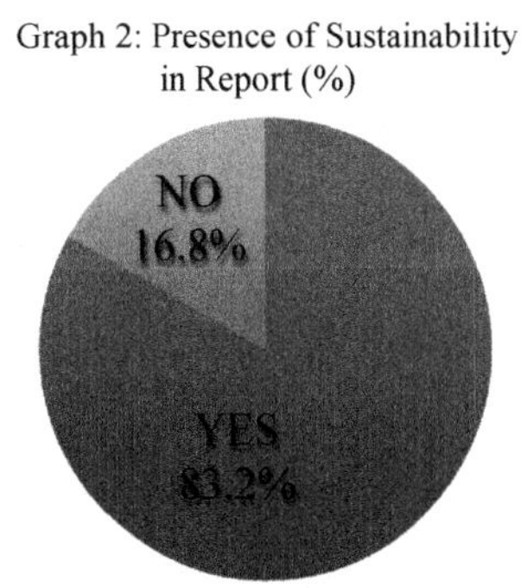

Fig. 2 Presence of sustainability in report (%)

Although more than 80 % of companies in this study included a reference to sustainability in their reports, only 60 % formulated sustainable development goals (Fig. 3). A slight majority of the sample created goals for their companies to approach sustainability in a more strategic and systematic manner.

The majority of companies included in this study lacked a strategic approach to sustainability, failing to include a time frame, as well as to state their SDGs in measurable terms (Fig. 4).

RQ2: How are SDG initiatives reported in terms of accountability to stakeholders?

Social initiatives are the most widely reported criteria among the three types of sustainable initiatives (Fig. 5). With companies focusing on mainly social issues, they demonstrate an emphasis on embracing their social responsibilities in the communities in which they operate. Environmental initiatives come second with 72 % of companies reporting them. The least reported types of initiatives are economic ones.

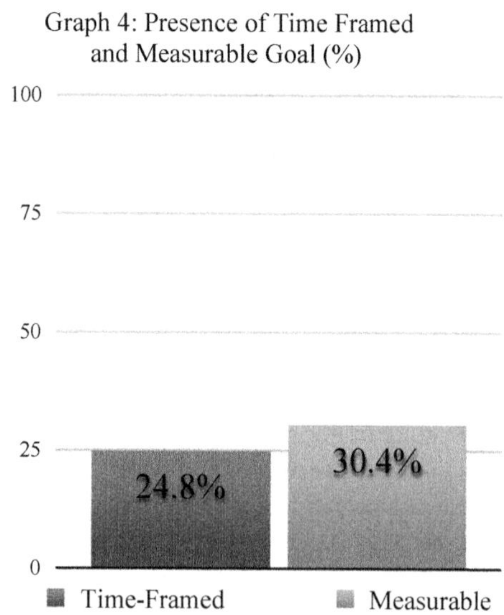

Fig. 3 Presence of formulated SD goals (%)

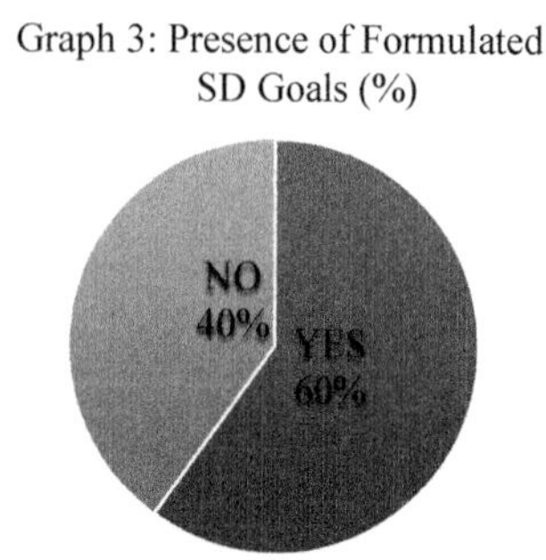

Fig. 4 Presence of time-framed and measurable goals (%)

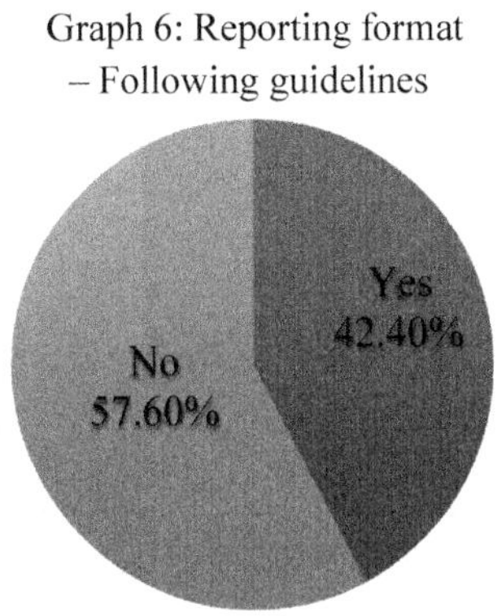

Fig. 5 Types of SDG initiatives (%)

RQ3: How are SDG initiatives measured?

Despite multiple recommendations by international organizations for companies to be more systematic in their reporting by using standardized guidelines, only 42.4 % of companies in this study reported using some types of guidelines (Fig. 6). The most popular guideline referred to was the latest version of the Global Reporting Initiative (GRI).

RQ4: How successful are the companies at reaching their SDG?

Findings indicate that only approximately half of the companies under investigation evaluate whether they have reached their objectives by actually measuring them.

RQ5: How do the five countries compare to one another?

There are clear differences in how companies in the five countries under investigation report their sustainable goals. Australia leads the other four in terms of

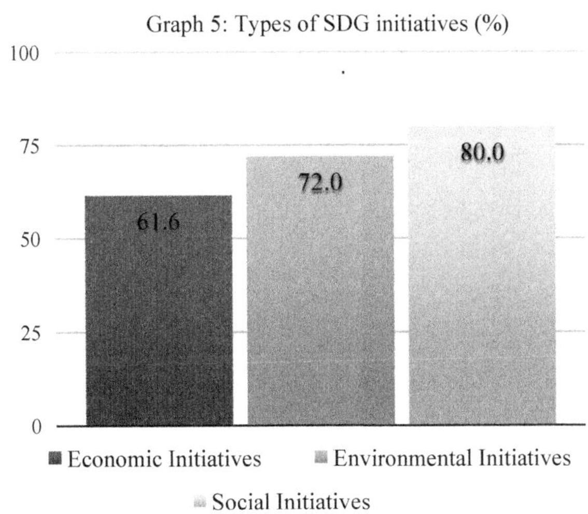

Fig. 6 Reporting format—following guidelines

Fig. 7 Evaluation of SDG (objectives vs. results)

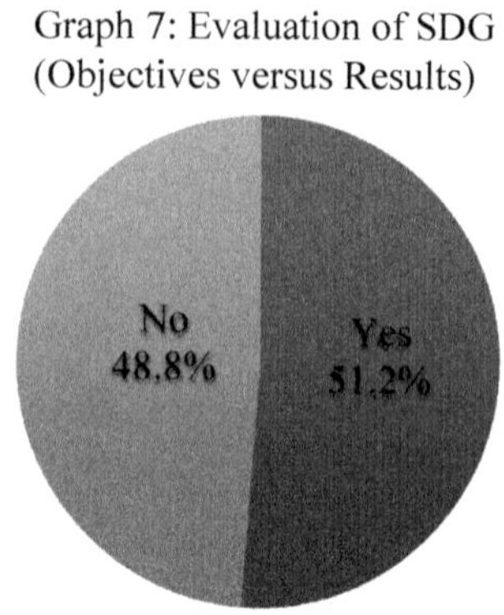

the number of companies reporting sustainable goals, as well as based on those including time-framed and measurable goals. Although Malaysian and Thai companies include sustainable goals in their reports, they lack specific goals that are time framed and measurable. Figure 8 indicates sample companies from Qatar and the UAE that are equally poor in regard to including sustainable, time-framed and measurable goals (Fig. 7).

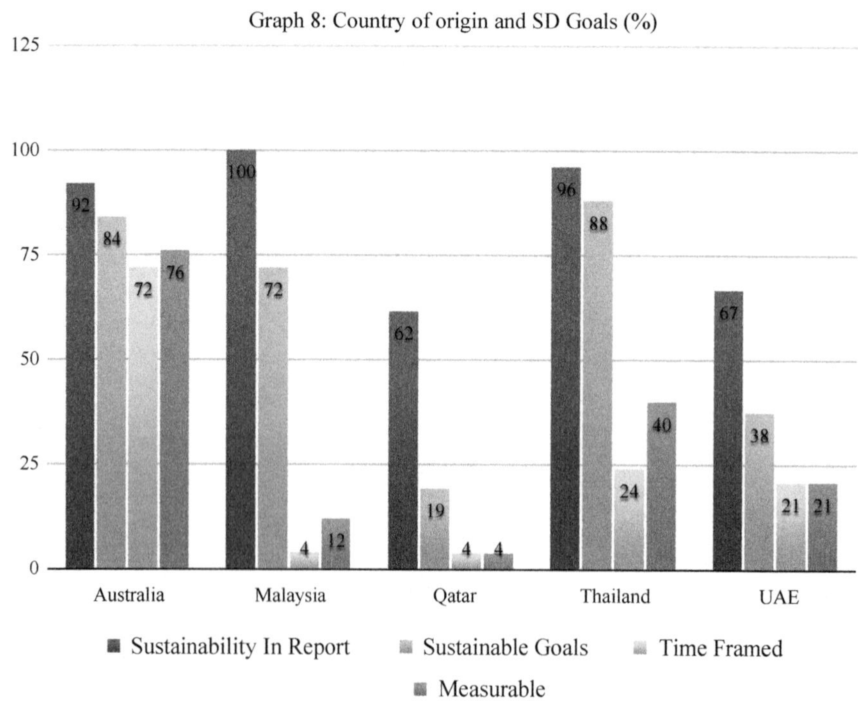

Fig. 8 Country of origin and SD goals (%)

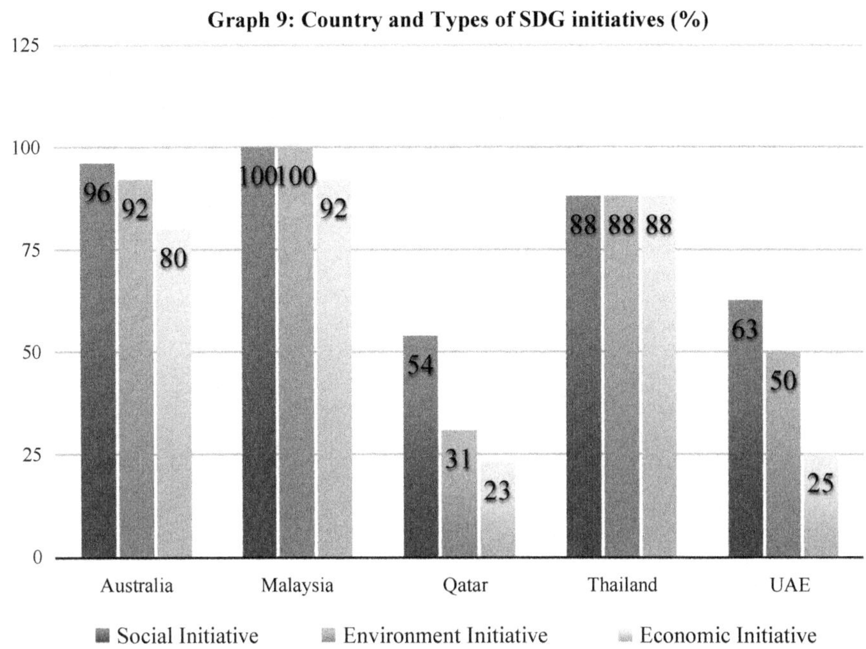

Fig. 9 Country and types of SDG initiatives (%)

All countries under investigation report engaging in all three types of initiatives, but to different degrees. Major companies from all five countries reported the most on social initiatives. Qatar and the UAE reported the least amount of all three types of initiatives but nevertheless predominantly reported on social initiatives (see Figs. 9 and 10).

Australia and Thailand are leaders in terms of following reporting guidelines with 80 % and 68 %, respectively, using either GRI, United Nations' Environment Programme (UNEP) or Dow Jones index reporting guidelines in their reports. Malaysia, Qatar and the UAE are lagging behind their regional neighbours in that respect.

Not surprisingly, based on the earlier findings, Qatar and the UAE lag behind Australia, Thailand and Malaysia in the measurement of their objectives with only 15.4 % and 20.8 %, respectively (see Fig. 11).

6 Discussion

The starting point for any discussion about sustainable development should be the acceptance of the priori fact that commercial organizations are rational agents. Thus, sustainability should be viewed as a pragmatic rational strategy, in which

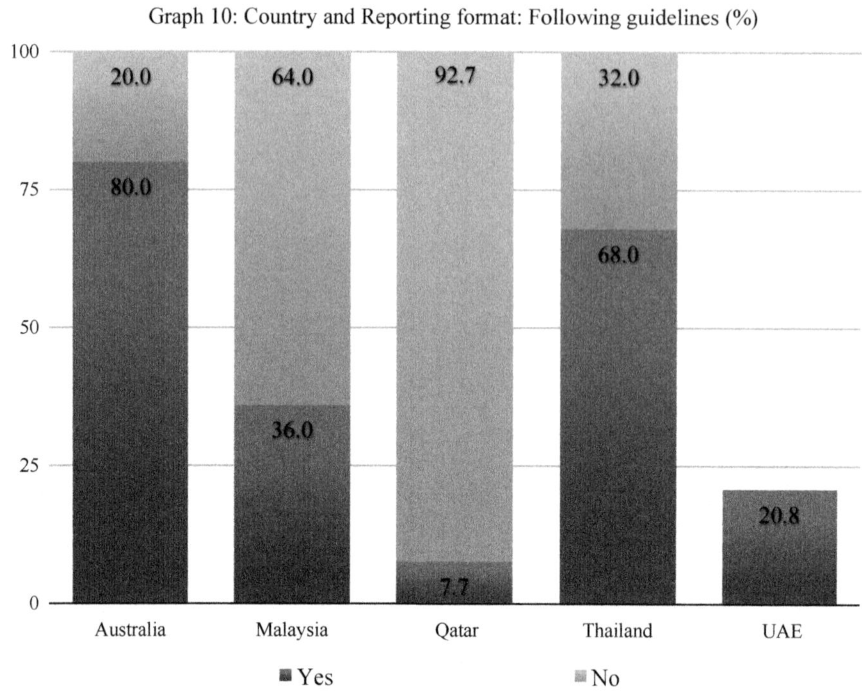

Fig. 10 Country and reporting format: following guidelines (%)

practical motives and results always stand behind the rhetoric. These motives may include image and reputation management, government relations facilitation and strategic long-term risk minimization (Garriga and Mele 2004).

These motives drive companies towards responsibility in the usage of their finite resources and the exploitation of their 'econiche'. This approach resonates increasingly with the public interest. It is where the most significant developments in measurability can be expected and where the internal pressure on organizations to be measurability driven seems to coincide with public demand to see sustainability in action.

It would thus be natural to expect companies to address their sustainability goals in public communication, and in particular in sustainability reports. The elements of sustainable development programs (social, economic, environmental) are supposed to have clear goals, and yet our analysis shows that 40 % of the companies under investigation fail to mention sustainability goals in their reports. In the absence of formulated goals, any evaluation becomes impossible—and sustainability programs risk shifting from being a core managerial activity to becoming a competition in sophistry.

Seen as a management effort, sustainable development should become subject to measurability requirements. Measurability is a crucial factor that provides

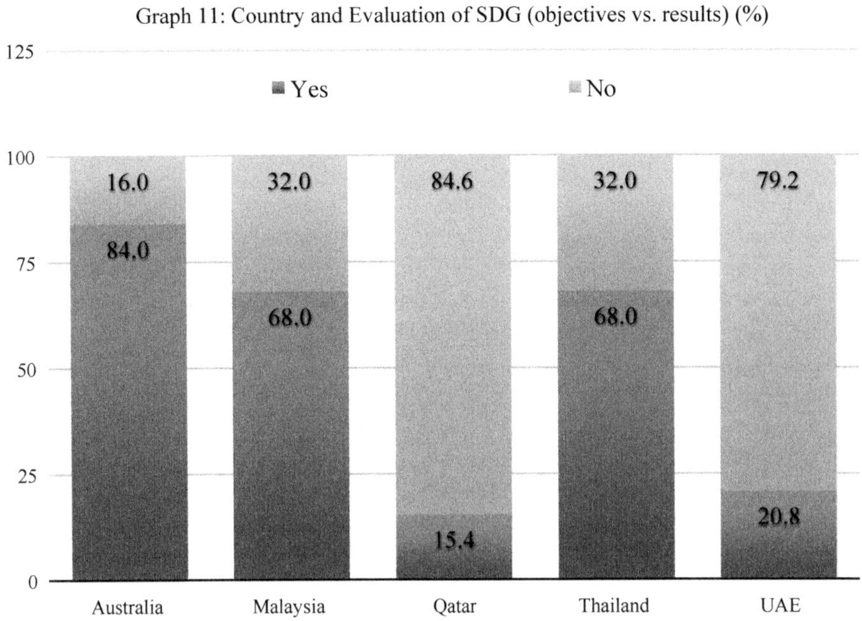

Fig. 11 Country and evaluation of SDG (objectives vs. results) (%)

theoretical and informational foundations for decision-making, effectiveness and efficiency of any managerial action depending on the quality and professionalism of measurement and evaluation. Thus, any sustainable development initiative must be examined through a rigorous evaluation lens. Sustainable development strategy implementation is more likely to be successful if based not on intuition and insights but on data, facts and empirical evidence. Rationality, measurability, business orientation, specificity and availability to evaluation are the key parameters that any managerial activity (including sustainable development efforts) must meet.

However, the crucial elements of business-related goals—being measurable and time bound—are not reflected in the majority of examined reports. This appears to contradict the declaration of sustainable development initiatives as parts of a business activity; in business, organizational long-term goals need to be clearly formulated to ensure a rational strategy for achieving them can be developed. This is a standard requirement for any business activity, and managers are usually reluctant to make any exceptions. If sustainable development is considered important, then making an exception for it would be inappropriate. On the contrary, the very assumption that sustainable development is important for business should set even higher standards for managerial planning towards measurable goals. However, although the lack of measurable and time-bound goals may highlight managerial shortcomings, they equally ensure less scrutiny and arguably accountability, due to limited transparency. It may not be best practice, but some organizations may therefore be inclined to keep their narratives broad—and reporting vague.

The guidelines by the Global Reporting Initiative (GRI), now in their fourth version, provide valuable guidance, insight and consistency in reporting beyond national and industry-specific requirements, rendering beneficial transparency and a unique opportunity for organizations to benchmark their performance. There are currently 180 laws and regulatory standards in 45 countries calling for corporate sustainability reporting. However, it is worth noting that according to the sample in this study the GRI reporting guidelines primarily appear to encourage to report but not necessarily to set organization-specific, measurable targets and to evaluate against these. 42.4 % of organizations in this study utilized the GRI guidelines to provide a snapshot of their current sustainability performance. However, some did so without putting their sustainability results into the context of their previous performance and long-term goals, hence limiting the value of sustainability reporting. Other guidelines commonly mentioned and referred to include the Dow Jones Sustainability Indices, as well as the United Nations Global Impact measures.

Based on this study, there appears to be two key drivers for comprehensive sustainability reporting and goal setting. First, imposed, national reporting requirements (and arguably to a certain extent action on sustainable goals and initiatives) require organizations to publicly acknowledge and report on their efforts over the past year. This can arguably be seen to a minor extent in Qatar, where 'Law No 13 of 2008' legally requires all organizations to contribute 2.5 % of their annual net consolidated profit to social and sports funds (in many cases, this was the full extent of all sustainability-related activity). Many businesses also referred to the Qatar National Vision 2030, influencing their business practices and goal setting. However, the standout example is Thailand, where all listed companies are required to demonstrate, in their annual registration statement and annual reports, how they apply 15 principles of good corporate governance. Furthermore, the Thai Securities and Exchange Commission (SEC), in conjunction with the CSR Club (under the Stock Exchange of Thailand), has requirements for listed companies to disclose their CSR (corporate social responsibility) operations in their annual reports, or in stand-alone reports. Consequently, 96 % of Thai reports analyzed in this study reported on sustainability initiatives; 88 % included sustainable goals, the highest country-based rate in this study. Reports furthermore included frequent references to the CSR Club and sustainability-related, national awards. Remarkably, even though Malaysia embarked on corporate responsibility and governance reporting initiatives slightly earlier, Thailand's quality of corporate sustainability reporting has outpaced its neighbour.

The second driver appears to be the industry an organization is operating in, leading to possible additional reporting requirements (for example, although sustainability reporting is currently voluntary in Australia, providers of financial products with an investment component are required to disclose the extent to which labour standards or environmental, social or ethical considerations are taken into account in investment decision-making (Parliament of Australia 2010). Further, drivers are community expectations as well as 'opportunities to report' and evaluate any impact. For example, businesses in the transport or resources sectors inevitably have a greater impact on environmental resources and/or pollution, in contrast to, for example, banking, where sustainability decisions are largely focused on diversity and investment decisions (i.e. sustainability criteria for investment decisions). Hence, community expectations may be greater, in addition to levels of scrutiny and even legal obligations to limit environmental and social impacts. Equally, businesses in mining, production, construction and transport inevitably need to pay greater attention to workplace safety, which again is reflected in the report samples, in terms of references to injury rates and similar. While globalization is regarded as a key driver for sustainability, it is evident that many companies still continue to implement sustainable practices comfortably from primarily legal and regulatory realms (Higgins and Debroux 2009). Overall, environmental and societal bottom line reporting remains limited and prosaic, supporting earlier studies that companies still lack understanding and appreciation of disclosures on impact of business on biodiversity and its management (Sharma 2013). This advertently exhibits the lack of foresight on the importance of sustainable development initiatives that contribute towards a healthier business ecosystem in the long run.

Consequently, besides the varying, national requirements observed in this study, the type of dominant industries in the top 25 companies whose reports were analyzed in this study ultimately drives country trends. For example, resources and utility organizations are more broadly represented in Australia (16 % of sample), which arguably might explain why the inclusion of sustainability goals is second highest after Thailand (84 % compared to 88 % in Thailand), despite the lack of any reporting requirements. Many of the organizations included are multinational and are familiar with (global) reporting expectations. The Australian sample included the highest rate of measurable (76 %) and time-bound (72 %) sustainability goals, as well as the highest rate of organizations who systematically measured their sustainability objectives (84 %). The Australian sample also contained the highest reference rate to sustainability guidelines (80 %) in this study. In contrast, banking and financial services are the most represented industries in Qatar (32 %) and the UAE (52 %), which may explain why the inclusion of sustainable goals was comparatively low (19.2 and 37.5, respectively), with only 3.8 % of time-framed and/or measurable goals in Qatar (compared to 20.8 %) and in the UAE. Here, a strong focus was placed on sustainable growth and return on investment (ROI) for shareholders (ironically, details on economic initiatives were only included in 23.1 and 25 % of reports, respectively, the lowest rate of any country sample included in this study). In the case of Qatar community, expectations most likely influenced the level of reporting, limiting most organizations' sustainability-related activities to the legally required funding of social and sports funds. Tellingly, only two organizations (7.7 %) included a reference to sustainability guidelines in their reports. Qatari organizations also provided the lowest rate of dedicated sustainability & CSR reports (8 %, n = 2), compared to 68 % in Thailand, 48 % in Australia and 44 % in Malaysia.

Other country-specific foci worth mentioning are the emphasis on diversity (in particular, gender equality and indigenous representation) in Australia. Malaysian companies favoured case study style formats, providing in-depth insights into sporting achievements, scholarship winners and individual successes, as well as social media activities, all of which barely featured in the other reports analyzed. The Malaysian sample also highlighted green buildings and developments.

Effectiveness of sustainable development of business is what gives justification of expenses for the business itself. However, the problem of sustainable development evaluation and measurement remains one of the important obstacles on the way for further development of sustainability initiatives. This problem persists at all stages: from planning of particular programs and projects, to justifying and defending them, to eventually evaluating the results. The situation becomes more complicated due to the variety of forms of sustainable development initiatives, projects and strategies, as well as the general problem of effectiveness.

7 Limitations

As the aim of this study was to compare the depth and breadth of reporting of sustainability development goals across selected countries in the wider Asian region, the research team identified a yes/no coding scheme as the most suitable approach for the desired comparison across the 12 coding items. However, it is worth noting that this simplified approach fails to accurately reflect the wide range —and depth—of reporting present. For example, some organizations provided in-depth information into how their sustainability efforts, across all categories (i.e. environment, social and economic), has evolved over time, while others only included easily accessible snapshots (most commonly on financial data). Equally, the quality of sustainability goals varied greatly. 60 % of reports analyzed included sustainable goals. However, only a minority included specific, time-framed (24.8 %) and measurable (30.4 %) sustainable goals. The remaining reports included statements with varying degrees of vagueness. The Malaysian sample illustrates this point: All reports in this sample covered sustainability, 72 % of which contained sustainable goals. However, only 4 % of these were time framed and 12 % measurable, indicating the ambiguity of many of the sustainability goals set, thereby limiting transparency and opportunities for benchmarking.

Overall, there has been major progress in terms of corporate sustainability reporting, but there are equally many ways in which sustainability reporting can be further improved. Most companies are not using measurable and time-framed objectives despite engaging in many economical, environmental and social sustainability initiatives. Further studies, including a broader sample of countries and a more complex coding system, which captures those nuances in reporting styles discussed above, are needed to further improve our understanding of how sustainability reporting is evolving and does vary between different economies in the wider Asian region.

8 Conclusion

On the whole, most emerging economies in Asia exhibit a myopic tendency towards sustainability and appear to be satisfied with 'meeting basic requirements set by relevant authorities'. Interestingly, companies in emerging economies with a multinational presence and an executive management with a strategic global market outlook outpace the competitors as they break the norms of corporate sustainability reporting by setting new frontiers in sustainable practices fortified with affirmative time-framed data and targets. As expected, companies in a fully developed economy context perform far better from a reporting perspective. Under further scrutiny, it becomes evident that corporate attitude in developed economies is spurred by broader-based risks, for example, stakeholder pressures that directly impact global brand value and corporate reputation, compared to companies operating in emerging markets—as many have yet to establish as global brands.

In addressing the current lag in corporate sustainability practices, it is pertinent to consider a multi-prong strategy. In Asia, apart from government intervention, a crucial dimension which aligns with Hofstede's (2005) high-powered distance cultures, there is strong need for wider stakeholder involvement from industry associations, NGO, multinationals, quasi-government regulatory institutions, as well agencies to expedite the necessary corporate attitude transformation in order to realize the SDGs' 2030 target. Next, a culture of compliance to legislative and regulatory frameworks should be enforced among companies. In most cases, it is not a lack of legislation, but rather the lack of enforcement that repudiates reporting standards. A well-planned multi-stakeholder network strategy should be mobilized by governments in the respective industries to strengthen the compliance domain towards sustainable practices. Over time, a compliance-driven approach would push companies to become more mindful towards a sustainability development discourse within their own country, as well as part of the global marketplace. Thirdly, companies in Asia need to move beyond the traditional philanthropic mantra towards communities and embrace affirmative commitment to both community and the environment by setting explicit sustainable targets that measure year-to-year corporate sustainability performance. The strong rationale is that post-2015 SDGs' launch presents an opportune moment for companies to strategize on how to set themselves apart in the face of intense competition, as new players from developing regions join the market, as well as survive despite a glum global economic outlook. While the awareness of SDGs may be high among commercial and public sectors, not all companies are clear on the course of tangible action for business advantage. By aligning corporate sustainability initiatives with Sustainable Development Goals (SDGs), companies create opportunities. Here, the ability to widen one's business attracts premium investors with superior market access, rather than merely settling with importers or business partners with narrow returns. Poor returns often lead to the spiralling down of business performance as management gears towards cost-saving measures—often followed by deprived fair labour practices, poor supply chain management, inevitable social media backlash and susceptibility to depleted corporate reputation.

References

Association for Sustainable Development in Asia (AsrIA). (2015). The State of Sustainable and Responsible Investment in Asia. *Green Money Journal*. Downloaded on 01st March 2016 from http://www.greenmoneyjournal.com/July-2015/asia/

Baeten, G. (2000). The tragedy of the highway: Empowerment, disempowerment and the politics of sustainability discourse and practices. *European Planning Studies, 8*, 69–86.

Baskin, J. (2006). Corporate responsibility in emerging markets. *The Journal of Corporate Citizenship, 24*, 29–47.

Bernstein, S. (2002). Liberal environmentalism and global environmental governance. *Global Environmental Politics, 2*(3), 1–16.

Bloomberg. (2015). *The world's 20 largest economies* in 2030. Accessed on 19 September 2015 at http://www.bloomberg.com/news/articles/2015-04-10/the-world-s-20-largest-economies-in-2030

Bloomberg. (2014). *Qatar, China Top Ranking of Frontier and Emerging Markets*. Contributed by Weiyi Lim on 29 January 2014. http://www.bloomberg.com/news/articles/2014-01-29/qatar-china-top-ranking-of-frontier-and-emerging-markets

Bloomberg Business. (2013). The top 20 emerging markets accessed on 19 September 2015 from http://www.bloomberg.com/slideshow/2013-01-30/the-top-20-emerging-markets.html

Brundtland Report. (1987). *Our common Future*. World Commission on Environment and Development. Oxford: Oxford University Press.

Chambers, E. W., Chapple, W., Moon, J., & Sullivan, M. (2003). CSR in Asia: A seven country study of CSR website reporting. *ICCSR Research Papers*. No 10.

Chapple, W., & Moon, J. (2005). Corporate Social Responsibility (CSR) in Asia: A seven-country study of CSR website reporting. *Business and Society, 44*(4), 415–441.

Chartterji, A. K., & Toffel, M. V. (2016). Do CEO activists make a difference? Evidence from a field experiment. Working Paper 16-100. 01st March 2016. Harvard Business School. Accessed from http://ssrn.com/abstract=2742209

Chen, S., & Bouvain, P. (2009). Is corporate responsibility converging? A comparison of corporate responsibility reporting in the USA, UK, Australia and Germany. *Journal of Business Ethics, 87*, 299–317.

Demeritt, D. (2006). Science studies, climate change and the prospects for constructivist critique. *Economy and Society, 35*, 453–479.

Dodd, M. (2016). *Corporate activism: The new challnege for an age-old question*. Downloaded on 28th April 2016 from http://www.instituteforpr.org/corporate-activism-new-challenge-age-old-question/

Dovers, S. (1996). Sustainability: Demands on policy. *Journal of Public Policy., 16*, 303–318.

Fifka, M. S. (2013). Corporate responsibility reporting and its determinants in comparative perspective—a review of the empirical literature and a meta-analysis. *Business Strategy and the Environment, 22*, 1–35.

Fortune. (2015). *The new world of Business*. Contributed by Ian Bremmer (February Issue 2015). Accessed on 20th September 2015 from http://fortune.com/2015/01/22/the-new-world-of-business/

Garigga, E., & Mele, D. (2004). Corporate social responsibility theories: Mapping the territory. *Journal of Business Ethics, 53*, 51–71.

Gray, S. J., Radebaugh, L. H., & Roberts, C. B. (1990). International perceptions of cost constraints on voluntary information disclosures: A comparative study of UK and US multinationals. *Journal of International Business Studies, 21*(4), 597–622.

Hartman, L. P., Rubin, R. S., & Dhanda, K. K. (2007). The communication of corporate social responsibility: United States and European Union multinational corporations. *Journal of Business Ethics, 74*(4), 373–389.

Higgins, C., & Debroux, P. (2009). Globalization and CSR in Asia. *Asian Business and Management, 8*, 125–127. doi:10.1057/abm.2009.3.

Higgins, C., Milne, M. J., & van Gramberg, B. J. (2015). The uptake of sustainability reporting in Australia. *Journal of Business Ethics. 129*(2), 445–468.

Hofstede, G., & Hofstede, G. J. (2005). *Cultures and organizations*. McGraw Hill, London: Software of the mind.

Holland, L., & Foo, Y. B. (2003). Differences in environmental reporting practices in the UK and the US: The legal and regulatory context. *The British Accounting Review., 35*(1), 1–18.

International Union for the Conservation of Nature and Natural Resources, World Conservation Strategy: Living Resource Conservation for Sustainable Development (1980). Available at http://data.iucn.org/dbtw-wpd/edocs/WCS-004.pdf

Kolk, A., & Perego, P. (2010). Determinants of the adoption of sustainability assurance statements: An international investigation. *Business Strategy and the Environment, 19*, 182–198.

Kolk, A. (2008). Sustainability, accountability and corporate governance: Exploring multinationals' reporting practices. *Business Strategy and the Environment, 17*(1), 1–15.

KPMG International. (2013). The KPMG survey of corporate responsibility reporting 2013. Retrieved from https://www.kpmg.com/Global/en/IssuesAndInsights/ArticlesPublications/corporate-responsibility/Documents/kpmg-survey-of-corporate-responsibility-reporting-2013.pdf

Maignan, I., & Ralston, D. A. (2002). Corporate social responsibility in Europe and the US: Insights from businesses' Self-presentations. *Journal of International Business Studies, 33*(3), 497–514.

Morhardt, J. E. (2010). Corporate social responsibility and sustainability reporting on the internet. *Business Strategy and the Environment, 19*, 436–452.

MSCI Emerging Market Index. (2015). Downloaded from https://www.msci.com/emerging-markets Accessed on 15 September 2015

Newson, M., & Deegan, C. (2002). Global expectations and their association with corporate social disclosure practices in Australia, Singapore and South Korea. *The International Journal of Accounting, 37*(2), 183–213.

Organisation for Economic Co-operation and Development OECD. (2001). Sustainable development: Critical issues. A Policy Brief published in the OECD Observer 2001. 1–8 downloaded from http://www.oecd.org/dataoecd/29/9/1890501.pdf

Packing a mightier punch: Asia's economic growth among global markets continues Asia Pacific Economic Outlook, Q1. (2016). by Akrur Barua. Deloitte University Press. Downloaded on 10 April 2016 from http://dupress.com/articles/asia-pacific-economic-outlook-q1-2016-asia-economic-growth-continues/

Parliament of Australia. (2010). *Sustainability reporting*. Retrieved 20th September, 2015, from http://www.aph.gov.au/About_Parliament/Parliamentary_Departments/Parliamentary_Library/Browse_by_Topic/ClimateChange/responses/economic/Sustainability

Porter, M. E., & Kramer, M. R. (2006). Strategy and society: The link between competitive advantage and corporate social responsibility. *Harvard Business Review, 84*(12), 56–68.

Sharma, B. (2013). *Contextualising CSR in Asia: Corporate social responsibility in Asian economies*. Singapore: Lien Centre for Social Innovation.

Smith, J. V. D. L., Adhikari, A., & Tondkar, R. H. (2005). Exploring differences in social disclosures internationally: A stakeholder perspective. *Journal of Accounting and Public Policy, 24*(2), 123–151.

Svensson, G., & Wood, G. (2011). A conceptual framework of corporate and business ethics across organizations: Structures, processes and performance. *The Learning Organization, 18* (1), 21–35.

Tsang, S., Welford, R., & Brown, M. (2009). Reporting on community investment. *Corporate Social Responsibility and Environmental Management, 16*, 123–136.

UN. (2014). Country classification http://www.un.org/en/development/desa/policy/wesp/wesp_current/2014wesp_country_classification.pdf

UN Global Compact. (2016). Downloaded on 22 April 2016 https://www.unglobalcompact.org/participation/join

United Nations. (2015). Sustainable development goals: 17 goals to transform our world. Downloaded on 15 March 2016 from http://www.un.org/sustainabledevelopment/
Welford, R. (2004). Corporate social responsibility in europe. North America and Asia. *Journal of Corporate Citizenship*, *17*: 33–52.
Willard, B. (2007). The NEXT Sustainability Wave. *The Malaysian Accountant Magazine*, February Issue, Malaysia.
Williams, S.M. (1999). Voluntary environmental and social disclosure practices in the Asia-Pacific Region: An international empirical test of political economy theory. *The International Journal of Accounting* 34(2): 209–238.
World Bank. (2015). Country and lending groups. Accessed on 20 September 2015 at http://data.worldbank.org/about/country-and-lending-groups
World Business Council for Sustainable Development (WBCSD). (2015). Business action on sustainable development goals—Paris 2015. Downloaded on 15 March 2016 from www.wbcsdevents.org/Paris2015/portfolio-item/business-action-on-sustainable-development-goals-sdgs/

Information Communication Technologies (ICT) for Education Projects in ASEAN: Can We Close the Digital Divide?

Pornpun Prajaknate

Abstract One of the main challenges facing the Association of Southeast Asian Nations (ASEAN) is to achieve sustainable development by bridging the digital divide throughout the region. During the past ten years, the member states of ASEAN have made significant investments in ICT infrastructure in order to increase ICT accessibility and adoption rates among the people residing in the ASEAN region. ASEAN, through the establishment of the ASEAN Economic Community (AEC), launched the ASEAN ICT Master Plan 2015 in 2011. The sixth strategy of this plan is "bridging the digital divide" in order to eliminate the ICT development gap across the ASEAN region and to increase ICT in education through various initiatives. This chapter synthesizes the qualitative and quantitative research articles published between 2005 and 2015 that focused on the implementation of ICT in the education programs in ASEAN. In light of Van Dijk's (The deepening divide: Inequality in the information society. Sage Publications, Thousand Oaks, Van Dijk 2005) causal and sequential model of digital technology, the objectives of this chapter are to provide a comprehensive overview of the current situation regarding ICT for education programs in ASEAN countries, and to identify the barriers to ICT adoption and use in education. The results indicate that some ASEAN member states such as Cambodia, Lao PDR, Myanmar, and Vietnam began investing in ICT infrastructures and Internet connection only in the last few years. On the other hand, other member states such as Thailand, the Philippines, Malaysia, Indonesia, and Singapore have further developed ICT projects in their schools. The teacher's technical mastery of ICT skills in integrating ICT into student learning in Thailand is insufficient. Teachers have suggested that the ICT training courses should include creating educational media lessons in order to enhance teachers' operational ICT skills. Accessing ICT in the Bruneian, Singaporean, and Malaysian schools is limited, despite much investment in ICT infrastructure, computers, and professional training. Barriers to such access still exist, and especially after receiving ICT training, teachers report lack of time to

P. Prajaknate (✉)
National Institute of Development Administration, Bangkok 10240, Thailand
e-mail: pornpun2@hotmail.com

J. Servaes (ed.), *Sustainable Development Goals in the Asian Context*,
Communication, Culture and Change in Asia 2,
DOI 10.1007/978-981-10-2815-1_6

integrate ICT into their classes due to heavy workloads. The findings offer policy makers ideas concerning guidance in terms of strengthening the efficacy of ICT for education programs and achieving the ASEAN ICT master plan 2015 in closing the digital divide and reaching sustainable development goals in the ASEAN region.

Keywords Information communication technology · Digital divide · ASEAN · Education program · Sustainable development

1 Introduction

Information communication technologies (ICT) play a crucial role in driving global development in various aspects, such as social, economic, and educational development (Peña-López 2015; Servaes 2014). However, the digital divide, which is a gap between those that have access to ICT and those that do not, still exists. According to the Global Information Technology Report 2015, the disparities in access are driven by the development of ICT. Despite improving Internet access, only 39 % of the global population have access to the Internet and about 10 % of the population living in low-income countries have Internet access. The high-speed broadband system, mobile application usage, and advance data services have progressed differently in poor and rich economic countries (Dutta et al. 2015).

The digital divide between regions remains enormous. The International Telecommunication Union (ITU) has stated that the percentage of people that use Internet is highest in Europe (77.6 %), followed by 66.0 % of the people living in the Americas region. The second last number of people that use the Internet are currently living in ASIA and the Pacific region, accounting for only 36.9 % (ITU 2015). In particular, several countries in the ASIAN region have been facing digital divide problems (Ness and Lin 2015). This is because the ASEAN region is highly diverse in terms of geography, population, and economic and social development (Beeson and Stubbs 2012). According to World Economic Forum estimates, the majority of ASEAN member states rank at the bottom half of the ICT readiness index, indicating the low rate of ICT affordability, skills and infrastructures; however, paradoxically, Singapore has the highest ICT readiness index and took the top rank in the world in 2015 (Dutta et al. 2015).

As the Association of Southeast Asian Nations (ASEAN) will become a single community, the ASEAN Telecommunications and IT Ministers have implemented the ICT development for the past several years to assist economic growth. Therefore, one of the main challenges facing the ASEAN community is to achieve sustainable development by bridging the digital divide throughout the region. ASEAN, through the establishment of the ASEAN Economic Community (AEC), launched the ASEAN ICT Master Plan 2015 in 2011. The sixth strategy of this plan is "bridging the digital divide" in order to eliminate the ICT development gap across the ASEAN region and to increase ICT in education through various initiatives (the ASEAN Secretariat, 2011).

At the present time, almost all of the ASEAN member states have integrated ICT in national education as well as in their ICT plans. According to a report of *Information and Communication Technology in Education in Asia* (UNESCO Institute for Statistics 2014), about half of the ASEAN member states (including Malaysia, the Philippines, Singapore, and Cambodia) have a stand-alone ICT in education plan, whereas the other half (e.g., Indonesia, Myanmar, and Thailand) have integrated ICT in their national education plans or national ICT plans. The Lao People's Democratic Republic is in fact the only country without an ICT education plan.

Regarding the integration of ICT in the national education curriculum, almost half of the ASEAN member states—Indonesia, Malaysia, Thailand, and Singapore—include a course on basic computer skills within the curricula at the primary, lower-secondary, and upper-secondary level of education. The rest of the ASEAN member states have not integrated basic computer skill courses at the primary level. Lao PDR and the Philippines provide a basic computer skills course for students enrolled at lower- and upper-secondary levels while Myanmar and Cambodia only provide this course at upper-secondary levels (UNESCO Institute for Statistics 2014).

Currently, none of the ASEAN member states have been able to achieve a universal 1:1 learner-to-computer ratio (LCR). According to the 2012 data of International Standard Classification of Education (ISCED) (UNESCO Institute for Statistics 2012), the LCR in all ASEAN member states is greater than 1:1. In Indonesia and Cambodia, over 100 primary and secondary school students have only one shared computer. The LCR values are considerably high for the secondary level of education in Cambodia (>500:1) and the primary level of education in the Philippines (412:1). In contrast, computers are more available in Singapore and Malaysia, where ICT in education is fully integrated in the national ICT policies and where eight and four students, respectively, share a computer at primary and secondary levels of education in Singapore. Moreover, there are about 9 students per computer at the secondary education level and 17 students per computer at the primary education level in Malaysia.

Despite the high levels of investment in ICT to improve the pedagogical practice of schools, there are several questions that exist concerning the success of the ICT in education programs in developing countries (Dutta et al. 2015). The challenge to governments of ASEAN member states, therefore, is to achieve sustainable development by bridging the digital divide throughout the region.

The main purpose of this study is to synthesize the qualitative and quantitative research articles published between 2005 and 2015 that focused on the implementation of ICT in the education programs in ASEAN member states, including Brunei Darussalam, Cambodia, Indonesia, Lao PDR, Malaysia, Myanmar, the Philippines, Singapore, Thailand, and Vietnam. The objectives of this study are to provide a comprehensive overview of the current situation regarding ICT for education programs in ASEAN member states and to identify the barriers to the integration of ICT in education of ASEAN member states.

2 The Current Status of ICT in Education in ASEAN

In order to increase understanding of current status of ICT in education among ASEAN member states, this section provides an overview of current status of ICT in education in ASEAN: first ICT in education policies and plans, and then ICT in education infrastructures in ASEAN.

2.1 *ICT in Education Policies and Plans*

In recognition of the diversity of geography, demography, economy, education, and culture in the Southeast Asian region, the Southeast Asian Ministers of Education Organization (SEAMEO 2010) has classified ASEAN countries members under UNESCO's four stages model of ICT development in education. The model (Fig. 1) identifies four stages of approaches to ICT integration in schools along the continuum: emerging, applying, infusing, and transforming (Anderson 2010). The model is elaborated as follows:

Stage 1: Emerging. Schools just start providing computers and basic skills lessons for students. The majority of teachers use ICT for working performance.
Stage 2: Applying. National ICT policies have been formulated. Teachers start applying ICT in school curricula and management functions.
Stage 3: Infusing. ICT has been integrated into the school curriculum. ICT and Internet connections are provided in most of classrooms. ICT is used to support the

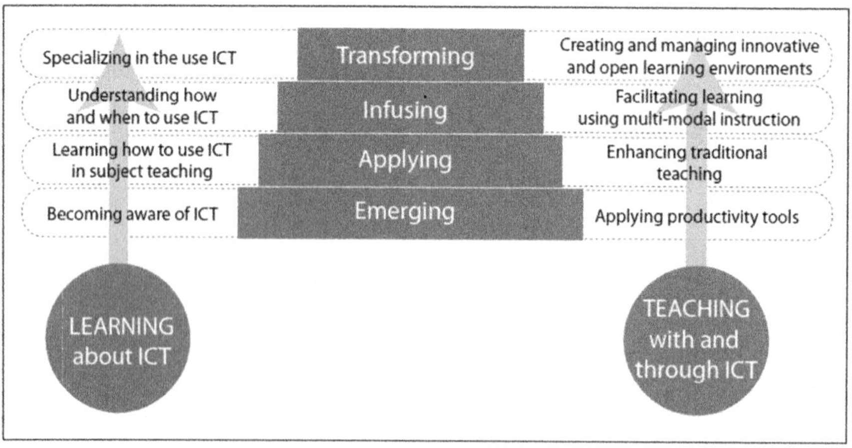

Fig. 1 Mapping stage for the integration of ICT in Schools. Adapted from Anderson (2010)

student project-based learning. The countries at the infusing stage have managed to regulate a national ICT plan and have established an ICT infrastructure. However, there is a wide disparity of ICT access and resources between rural and urban areas. *Stage 4: Transforming.* ICT is integral to the school system. Students are motivated to use ICT for innovative thinking and practice.

Regarding the policy-formulation process among ASEAN member states, the Lao PDR is listed at the emerging stage where ICT in education policies have not been launched yet (SEAMEO 2010). The priorities are placed on the establishment of ICT infrastructure, hardware, and software. The development of distance education is given a high priority in ICT policies in the Lao PDR (Baggaley and Belawati 2007). Cambodia and Myanmar are listed at the applying stage. ICT in education visions has been launched. The emphasis is placed up on the use of ICT for strengthening cultural and educational development (SEAMEO 2010). In Cambodia, the Ministry of Education, Youth, and Sport has launched the Education Strategic Plan, which shows high policy commitment to supporting investment in ICT infrastructures in schools at all levels (Kinuthia and Marshall 2010).

Indonesia, the Philippines, Thailand, and Vietnam are grouped at the infusing stage where ICT policies in education have been formulated (Clothey 2012). However, the gap between urban and rural areas is high. The ICT in education in rural areas lacks behind in teaching and learning pedagogies. Thailand and Vietnam show greater development of ICT in education than Indonesia or the Philippines (SEAMEO 2010).

Brunei Darussalam, Malaysia, and Singapore are categorized as transforming countries where ICT has been integrated into teaching, learning, and policies and has been used to enrich innovative ways of teaching and learning. The ICT policies in these three countries are integral parts of national ICT and education plans. Computers and other ICTs have been installed in almost all the classrooms with a high level of Internet connection (Symaco and Brock 2013). When comparing the ICT development in education between these three countries, Malaysia and Singapore have made greater advancement than Brunei Darussalam (SEAMEO 2010). For instance, Singapore has a decentralized ICT financial management system, with financial management activities performed in the schools (UNESCO Asia and Pacific Regional Bureau for Education 2004). As a consequence, the ICT programs devised from the schools' needs are tailored to suit the teachers' and students' learning environment. The "Smart Schools" project, an initiative of the Malaysian government. has successfully promoted using ICT in schools and in the community and has transformed the traditional school to the smart schools model in which students learn from a variety of ICT materials rather than teacher-to-student directed learning (Mondi et al. 2008). The "Smart School" project became a lessons-learned model to help guide ICT policy formulation and management of ICT in the school curricula of other countries (Kamel 2010) (Table 1).

Table 1 National policy in education plan by ASEAN member states

ASEAN member states	Policy/plan
Brunei Darussalam	MOE Strategic Plan 2007–2011 e-Brunei National IT Strategic Plan "IT 2000 and Beyond"
Cambodia	The Education Strategy Plan 2009–2013 (ESP) ICT-in-Education Master Plan 2009–2013
Indonesia	Five-Year Action Plan for the Development and Implementation of ICT in Indonesia
Lao PDR	National ICT Policies Education Sector Development Framework 2009–2015
Malaysia	Education Blueprint 2013–2025
Myanmar	ICT infrastructure development ICT Master Plan 2011–2015
The Philippines	Draft ICT4E strategic Plan (2008) and book version Strategic Roadmap for the ICT Sector (2006) National Framework Plan for ICTs in Basic Education (2004)
Singapore	Third Master Plan for ICT in Education (mp4) 2015–2019
Thailand	Eleventh National Economic and Social Development Plan 2012–2016 (NESDP) MICT Second National ICT Master Plan 2009–2013 (NICTMP) MOE Master plan on ICTs in Education 2007–2011(MICTE)
Vietnam	ICT in Education Plan (2001–2005)

Adapted from GeSCI. (2011). ICTs in Education (ICT4E): Global e-schools and Communication Initiative and UNESCO 2013 ICT in education, policy, infrastructure and ODA status in selected ASEAN countries. Bangkok: UNESCO Asia regional Bureau for Education

2.2 *ICT in Education Infrastructures in ASEAN*

The level of ICT infrastructure investment in education varies considerably between different ASEAN member states. Among the ten ASEAN member states, only the Lao PDR is at the emerging stage of ICT in education development. The basic infrastructures such as a national fiber optic backbone have been completely established throughout the country only with the last few years. By using the fiber optic backbone, seventeen model ICT centers have been recently built in 17 provinces (UNESCO 2013).

Cambodia and Myanmar have reached the applying stage, where the governments have put some efforts into ICT education projects. In Myanmar, the government has launched an e-Education project and the Ministry of Education has initiated e-Education learning centers using satellite data broadcasting and online training for an e-Learning system. As a consequence, a number of online training programs launched by universities and colleges have been provided at 200 learning centers (International Business Publications 2014). About 141 lower-secondary schools have built computer centers, and pre-service teacher training programs have been developed by regional teacher training colleges and the National Institute of Education (UNESCO 2013).

On the other hand, Thailand, the Philippines, Indonesia, and Vietnam, infusing-stage countries, have further developed ICT projects in their schools (Clothey 2012). For example, Thailand has implemented several ICTs in education projects including one tablet PC per child, the Thailand cyber university project, the ICT schools pilot project, Thailand SchoolNet, and distance education via satellite (UNESCO 2013). With the SchoolNet Thailand project and Thailand Cyber University project, students and teachers in several schools and universities are able to access information resources (Makaramani 2013; UNESCO 2013). Approximately 1000 online lessons for seven subjects have been formulated under the ICT schools pilot project (UNESCO 2013).

Singapore, Malaysia, and Brunei have reached the transforming stage at which ICT has become an integral part of the school curricula. For instance, the Malaysian government has invested RM6 billion for ICT projects during the past decades such as the Smart Schools Initiative. In order to ensure that all of the schools in Malaysia have comprehensively provided ICT infrastructure by 2015, 1BestariNet was launched in 10,000 schools (Malaysian Government 2012). In Singapore, personal infocomm devices have been used in education and learning to strengthen the engagement learning experience among students and teachers and to develop virtual classrooms. Students are able to obtain textbooks, lessons, and projects (iN2015 Steering Committee 2015). This is a school-based management program which promotes school self-initiative and autonomy. Each school must submit proposals in order to obtain funding for ICT development in the school projects from the Ministry of Education (International Business Publications 2012).

3 Methods

A systematic review of the research literature was undertaken in order to identify the barriers that affected the adoption and use of ICT in education programs in ASEAN member countries. This method enabled the researcher to comprehensively examine the multitude of studies with increased explanatory power (Brownson and Petitti 1998). Following the Causal and Sequential Model of Digital Technology formulated by, Van Dijk (2005) that identified four types of access-divide problems, a framing question was developed. In addition, a search for existent literature regarding ICT in education programs was performed.

Peer-reviewed articles published between 2005 and 2015 were systematically searched. The focus was on articles that had data on the use of ICT in education in ASEAN member states. A computerized search was conducted in Academic Search Complete, Communication and Mass Media, Education Research Complete, ISI Web of Science, JSTOR, Sage HSS Package, Science Direct, and Wiley-Blackwell Synergy using a combination of keywords related to ICT in education. The sets of key words included ICT in education, motivation access, material access, skill access, usage access, digital divide, and access problems and barriers.

Inclusion criteria included any articles that placed emphasis on ICT in education and the digital divide in Brunei Darussalam, Cambodia, Indonesia, Lao PDR, Malaysia, Myanmar, the Philippines, Singapore, Thailand, and Vietnam. With the data selection, all of the potential eligible full-text articles and abstracts were independently screened and assessed by two reviewers. Based up on the previous studies, the assessment of methodological quality was conducted using a 6-item scale. If disagreement occurred between two investigators, they attempted to reach a consensus through a discussion of differences. The investigators rated each item using the terms "yes," "no," or "unknown." Twenty-seven articles focusing on ICT in education conducted in ten ASEAN member states met the criteria (Appendix 1 and 2).

4 Literature Review

4.1 *Definition of Digital Divide*

The digital divide is not a single agreed-upon concept (Kariyawasam 2008). This section discusses different aspects of the digital divide: the traditional view, the availability of ICT, geographical dimension, and economic equality problems. According to the traditional view, the digital divide is described as "the gulf between those who have ready access to computers and the Internet, and those who do not" (Oxford Dictionaries 2014). From the perspective of the availability for use of information communications technology equipment and infrastructure, James (2003: 23) defines the digital divide as "the unequal distribution of computers, Internet connections and fax machines and so on between countries." He further describes the digital divide as "the strikingly differential extent to which rich and poor countries are enjoying the benefits of information technology."

Van Dijk (2005: 3) describes the digital divide as "a social and political problem, not a technological one. Physical access is portrayed as only one kind of (material) access among at least four: motivational, material, skills, and usage." From his point of view, the public and academic scholars unduly focus on the second stage. The majority of people think that increasing the amount of computer ownership and Internet access will eliminate the digital divide. Van Dijk argues against this notion, saying that attention should go beyond the provision of infrastructure because closing the material access gap alone is not enough to reduce the digital divide. He further indicated that it is important to motivate people to use ICT, enhance their digital skills in operating ICT, and provide them with enough time to use ICT.

Some scholars have emphasized the geographical dimension, focusing on the ICT disparity between domestic and international as well as urban and rural areas. Cammaerts and Van (2003) defined the digital divide as the "unequal access to technologies or digital exclusion at an international as well at a local level." Cullen (2003: 247) has stated that

> the digital divide has been applied to the gap that exists in most countries between those with ready access to the tools of information and communication technologies (ICTs), and those without such access or skills. This may be because of socioeconomic or geographical factors, educational, attitudinal and generational factors, or because of physical disabilities.

The digital divide can be defined from an economic perspective, demonstrating the inequality of access to ICT in terms of the socioeconomic differences between rich and poor. Sorj (2008: 65) addresses the digital divide as an economic inequality problem. From his perspective, the digital divide is a tool to eliminate poverty. He further elaborates that the digital divide demonstrates a huge disparity between the elites, who have the economic power to obtain ICT, and those that are economically disadvantaged, as follows:

> The initial social impact of the Internet did increase social inequality because it reached first the wealthiest sectors of the population. Thus, the fight against the digital divide is not so much a fight to diminish social inequality in itself as it is an effort to prevent inequality from increasing because of the advantages that those groups of the population with more resources and education enjoy a result of exclusive access to this information technology.

As can be seen above, these academic scholars discuss the digital divide in terms of technological and economic disadvantage. However, others have determined the digital divide from multi-dimensional aspects. Norris (2001: 4) examined the meaning of the digital divide according to three multidimensional aspects in the following:

> The global divide refers to the divergence of Internet access between industrialized and developing societies. The social divide concerns the gap between information rich and poor in each nation. And finally, within the online community, the democratic divide signifies the difference between those who do, and do not, use the panoply of digital resources to engage, mobilize, and participate in public life.

The Organization for Economic Co-operation and Development (Organization for Economic Co-operation and Development 2001: 5) examined the digital divide as "the gap between individuals, households, businesses and geographical areas at different socioeconomic levels with regard to their opportunities to access ICTs and to their use of the Internet for a wide variety of activities."

According to Wilson (2004: 300), the digital divide is "an inequality in access, distribution, and use of ICTs between two or more populations." Wilson identifies eight aspects related to the digital divide: physical access, financial access, cognitive access, design access, content access, production access, institutional access, and political access. In order to examine the level of accessibility to ICTs, Sorj (2008: 66) proposed five factors related to the digital divide: (1) the physical infrastructure, (2) the connection equipment, (3) ICTs training, (4) intellectual capabilities and social insertion, and (5) the production and use of specific contents.

From my perspective, it is impossible to identify a single comprehensive definition of the digital divide that is useful for every purpose; rather, it can be interpreted in different ways depending on the types of analysis and research purposes.

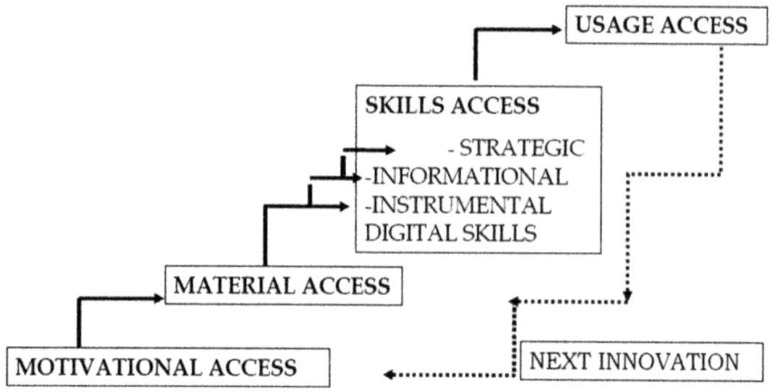

Fig. 2 The causal and sequential model of digital technology. Adapted from Van Dijk (2005)

4.2 Theoretical Framework

The causal and sequential model of digital technology developed by Van Dijk (2005, 2006) has been used as a framework to examine the inequality in the access to digital technology in the ASEAN region. The model classified four stages of access involved in the process of gaining access to the technology—from the motivation to use technology to the actual usage (Van Dijk 2009). The four stages include:

(1) motivational access (motivation to use digital technology),
(2) material access (possession of computers and Internet connections),
(3) skill access (possession of digital skills), and
(4) usage access (usage time) (Fig. 2).

5 Results

5.1 Motivation Access

Motivation is an initial stage of the whole process of access to technology devices and Internet connection. This stage involves individuals that do not have access (have-nots) and those that do not want to access the ICT (wants-nots) (Van Dijk 2005). The individuals' motivation for using ICT is influenced by several factors, including cultural, social, and psychological factors (Van Dijk 2009). The culture factor has an impact on the motivation to use ICT. Van Dijk has stated that men in masculine cultures refuse to handle the computer as it is considered something that women should do. In this case, men do not want to access the ICT. Regarding the social factor, Van Dijk (2005) described people that have a lack of opportunity to use computers as "truly unconnected." These groups consist of people with low income, low education, and the unemployed, such as elderly women and housewives.

Previous research conducted in Indonesia and Malaysia found the importance of the motivation factor in the use of ICT in schools. One path analysis study of 874 Indonesian teachers found that teachers' motivation plays a mediated role in the teacher's intention to integrate ICT into the teaching and learning in the classroom (Copriady 2014). In particular, Aziz and Shah (2015) conducted an in-depth interview among secondary school teachers in Malaysia to examine the barriers to the use of ICT in English language classrooms. The findings showed that older teachers have a lack of motivation to integrate ICT in the student learning processes because of their attitudes toward using ICT. Older teachers still hold a favorable attitude toward using traditional media in the classroom and a relatively negative attitude toward aging, stating that they are too old to attend training and to adopt new teaching methods.

Van Dijk (2006) further stated that psychological symptoms also impede the individuals' motivation to use ICT, such as technophobia (misgiving about the positive effect of ICT) and computer anxiety (fear and stress when using ICT). Previous research in Cambodia found that computer anxiety impedes the process of ICT teacher training. The Flemish Association for Development Cooperation and Technical Assistance has worked in close collaboration with the Teacher Training Department of the Ministry of Education, Youth and Sports on the integration of ICT in the teacher training centers of Cambodia by providing ICT teacher training and technology devices such as DVD players, television, projectors, and netbooks. However, the technology devices were not fully utilized because the teacher trainers, who are not digital natives, have felt some level of computer anxiety (Dionys 2012).

There is ample empirical evidence among ASEAN members states demonstrating that psychological factors hinders the students' and teachers' motivation in terms of using ICT in schools. In a quantitative study of 1137 students enrolled in secondary schools in Cambodia, Richardson et al. (2014) examined their perception toward the use of ICT. The results showed that students have a positive attitude toward technology, especially in terms of the opportunity to increase their computer competency. However, due to limited computer and Internet accessibility, poor ICT training support, and fear of damaging computers, the students have experienced computer anxiety and are reluctant to use computers.

In spite of the fact that Thailand launched the "One Tablet PC Per Child" (OTPC) policy in 2012, the digital divide still exists. About 800,000 tablets were distributed to first-grade students throughout the nation in 2012. However, a quantitative study undertaken among 213 second-grade students residing in the northern part of Thailand revealed that geographical factors and ICT use experience affected the students' acceptance of tablets. The students that lived in rural areas were more likely to have higher computer anxiety than those that lived in urban areas. Students less experience using tablets were more likely to have computer anxiety symptoms than students that had more experience (Pruet et al. 2014). In a survey of 500 adults students enrolled for undergraduate degrees at the University

Technology Mara in Malaysia, there was empirical evidence that adult students experienced poor confidence and computer anxiety when engaging in electronic distance learning (Abdullah et al. 2009).

5.2 Material Access

Material access has been defined as the ownership of ICT and network connection devices as well as an opportunity to access ICT and Internet at home, work, school, and with public networks (Van Dijk 2005). The existing literature on ICT use in education in the ASEAN region has demonstrated that inadequate access to ICT infrastructures, hardware, and software hinders the integration of ICT in the learning and teaching process in schools (Chan Yuen et al. 2011; Choy et al. 2009; Hamzah et al. 2009; Jayson 2011; Rodrigo 2005; Rye 2009; Shin et al. 2014; Tongkaw 2013).

Material access problems such as lack of infrastructure, poor Internet connections, and insufficient number of computers remain important issues for the applying stage of ICT for developing countries such as Cambodia and Myanmar. Shin et al. (2014) examined Cambodian teachers' perceptions and practices of integrating ICT in secondary school and explored the barriers to successful implementation. Questionnaires were distributed to 121 Cambodian teachers from secondary schools in Cambodia. The study found that although teachers had a positive attitude toward the use of ICT in the classroom, the actual use of ICT in the classroom was quite low. Fewer than 30 % of teachers reported using ICT devices in the classroom because of poor ICT infrastructures and Internet connections, insufficient ICT, and out-of-date computers. The results of face-to-face interviews with 17 Cambodian teacher trainers showed that the main barriers to the adoption of ICT, including poor ICT infrastructure and Internet connections, obsolete computer technology, and an insufficient number of computers, inhibited the use of ICT and the improvement of ICT teacher trainers' ICT skills (Jayson 2011).

Inequality of access to ICT material occurs during the infusing stage of ICT development in countries such as Indonesia, the Philippines, and Thailand. This has hampered the efforts to integrate ICT at all education levels among these countries.

At the primary and secondary education levels, Rodrigo (2005) determined the Internet access rate and student/computer ratio of schools in the Philippines, Thailand, Norway, Russian Federation, Slovenia, and South Africa based on the International Association for the Evaluation of Educational Achievement survey (IEA). Compared with other countries, the Internet access of schools in the Philippines was limited. Schools there have limited ICT infrastructures which prevented the students from accessing the digital world.

At the higher education level, the inequality of access to ICT infrastructure and low Internet connections still occur in the higher education institutions in the rural areas of Thailand and Indonesia. The integration of ICT in many universities has been impeded by the insufficient basic infrastructure (software and hardware), poor resources management, and maintenance plans, and poor provision of technical support. These conditions prevent teachers from making full use of ICT in the classroom (Tongkaw 2013). Rye (2009) found that poor ICT infrastructures and low Internet s have hindered students' access to ICT at the Universitas Terbuka located in southern part of Sumatra, Indonesia. These problems were one of the reasons for the students' resignation from the master program in public administration.

There is evidence that the digital divide still exists even in Singapore and Malaysia, which has reached the last stage, the transforming stage of ICT development in education. In Singapore, Choy et al. (2009) conducted pre- and posttest evaluation among 118 students enrolled in a postgraduate teacher education program in Singapore and participated in a technology training course. Their intention to use ICT for student-centered learning was quite high. However, the respondents failed to translate that intention into practice and reported difficulty integrating ICT in the teaching lessons because of the lack of updated software and low-speed Internet.

In Malaysia, Chan Yuen et al. (2011) examined the teachers' attitude, competency, and practice toward the integration of ICT in a teaching English as second language course in Malaysia. The results showed that the major barrier to integrating ICT in the classroom was a lack of ICT facilities and ICT technical service in the schools. Another qualitative study undertaken in four smart schools in the state of Negeri Sembilan and Selangor, Malaysia addressed the idea that an inadequate number of computers for teachers and students prevented teachers from integrating ICT in the teaching process. The computer-sharing rate was about 8 teachers per computer (Hamzah et al. 2009). In a pilot study that examined the computer ownership, skills, and usage of 30 part-time first-year students enrolled in a bachelor in education program at a local private university located in the state of Selangor, Malaysia, Kaur and Sidhu (2010) found that about 17 % of the respondents did not own computers and had difficulties in accessing the Internet; their Internet usage time was quite low. As can be seen from the results, the majority of respondents (80 %) spent fewer than 3 hours a week.

5.3 *Skill Access*

After motivating people to use ICT and to make a purchase, it is crucial to acquire the skills to effectively use ICT. Van Dijk emphasizes that the disparity in ICT skills access is much wider than motivational and material access (Van Dijk 2005). Skill access consists of four types of skills: operational skills (individual knows how to use ICT), formal skills (individual knows how to manage some functions of ICT such as hyperlinks and menus), information skills (individuals know how to search, choose and assess information in ICT), and strategic skills (individuals effectively use ICT to achieve professional goals) (van Deursen and van Dijk 2009).

Previous studies found that teachers working in ASEAN member states had low digital skills. A qualitative study examined the use of ICT in school counseling of 20 teachers residing in Luzon, the Philippines found that teachers with economic disadvantages and poor digital skills had insufficient access to the Internet (Masagca and Londerio 2008). In a quantitative study among 783 Vietnamese teachers, Peeraer and Van Petegem (2011) found that respondents reported a lack of computer confidence and more than half of the respondents had low basic computer, Internet, and maintenance and security skills. Safitry et al. (2015) examined the teachers' attitude and use of ICT in teaching and learning among 30 elementary school teachers with 10 years of teaching experience in Cipayung, Depok, West Java, Indonesia. About half of the respondents reported a low competency level in using Microsoft Office (40 %), search engines (50 %), and e-mail (50 %). Only 30 % of the teachers frequently used computers in the classroom while the rest of the respondents rarely and never used them.

Previous studies conducted among ASEAN member states have demonstrated teachers' emphasis on the importance of enhancing ICT skills that could help with the integration of ICT into the learning process as well as school management functions. Akarawang et al. (2015) identified the ICT training barriers and needs among teachers living in the northeastern part of Thailand. In spite of the fact that most Thai teachers have the ability to access ICT devices and connect to the Internet, the teacher's technical mastery of ICT skills in integrating ICT in student learning was found to be insufficient. Their requirements to strengthen ICT skills included training in general computer use, the Internet and networking, and Microsoft Office. Teachers have suggested that ICT training courses should include creating educational media lessons in order to enhance the teachers' operational ICT skills. Arokiasamy et al. (2015) conducted a quantitative research project among 520 secondary school principals in the state of Selangor and Wilayah Persejutuan, Malaysia, and found that the school principals had poor digital competencies in terms of their ability to obtain data from the Internet and to work on database spreadsheets and with multimedia software. Thus, these digital skills are crucial keys to becoming great technology leaders. Providing school principals with ICT development programs is critically needed for strengthening ICT professional skills in managing schools in this era of technology.

5.4 Usage Access

Usage access is actual usage, which is the last step to reach the ICT devices and network connections. Individuals may be encouraged to use and purchase ICT devices and Internet connection and to possess digital skills. However, the gap in the digital divide may occur within this step if individuals have no demand, opportunity, responsibility, time, or effort to use ICT devices. In particular, motivation and material access measure the possibility to use ICT and Internet connections. However, the usage access measures the actual use of ICT, which differs from the first three steps of access. For example, schools may provide computers with Internet connections for students, but it remains inadequate to meet the demand of students. Computers may be available at work but officers may be too busy to use them (Van Dijk 2005).

Regarding the inequality in usage access among ASEAN member states, Yieng and Saat (2013) investigated how Malaysian teachers integrate ICT in science lessons and the effects on students' performance using data from the trends in International Mathematics and Science Study 2011 (TIMSS). The results revealed that the computer usage rate per month was very low, accounting for 14–17 %, and the percentage of students that used computers in the science classroom was only 17 %. Students rarely searched or obtained information regarding scientific procedures or experiments on the Internet, which may have caused low science scores compared with other countries. Peeraer and Van Petegem (2011) found that Vienamese teachers have a higher computer usage rate for school administrative tasks rather than teaching tasks.

A country such as Singapore, where infrastructure availability and teacher–student digital skills are high, strict education policies and the teacher's heavy workload reduce the opportunity to use ICT resources among students and it remains a key development challenge. In a qualitative case study research, Lim (2007) examined the implementation of the Masterplan for ICT in Education (MP1) by integrating ICT in ten schools in Singapore in order to enhance the students' higher order thinking. The result showed that current school systems place tremendous pressure on teachers and students to succeed on the national examination. As a consequence, teachers and students prioritize support for "improving examination results" rather than teaching and learning with ICT in schools. Thus, the objective of strengthening student achievement could limit the opportunities for students and teachers to engage in ICT learning activities.

Heavy workloads also prevent teachers from using ICT in the learning and teaching process. The results from a qualitative study of the teacher working in five Singaporean schools that pioneered the "Future School" initiative in 2008 demonstrated that the teachers' intention and their practice in integrating ICT into teaching and learning demonstrated that a heavy workload is a major hindrance factor in the integration of ICT into the learning and teaching process. Moreover, the integration of ICT in school lessons is time-consuming. Therefore, teachers prefer using traditional methods rather than ICT to prepare their learning and teaching lessons (Lim 2015).

A number of previous studies undertaken among ASEAN member states have demonstrated that lack of time has hindered the use of ICT in the classrooms. Gardner and Yaacob (2009) found that lack of time prevented English language teachers in four primary schools in Kedah, Malaysia, from reviewing CD-ROMs and integrating them into their English language lessons. This is similar to the results from a qualitative study conducted among 21 secondary school teachers, where Wan Zah Wan et al. (2009) investigated the barriers to the integration of ICT in a smart school project launched by the Ministry of Education, Malaysia. The findings showed that insufficient time to prepare teaching lessons using ICT was reported to hinder the efforts to integrate ICT into the teaching and learning process. Teachers usually take time for obtaining information regarding their teaching lesson. In the mixed-method research of teacher educators participating in Technological Pedagogical Content Knowledge workshops in Vietnam, Peeraer and Van Petegem (2012) distributed a survey to 505 teachers at the beginning and the end of a workshop and conducted a focus group discussion among 67 teacher educators in order to assess the outcome on ICT professional development. The results at the end of the training showed that the teacher educators were confident in adopting ICT into their teaching lesson; however, some teacher educators expressed concern over the actual use of ICT in the classroom as they lacked time to practice and integrate ICT into their classroom.

Interestingly, uncommitted school leaders prevent teachers from using ICT in the classroom. Seyal (2012) conducted a survey among 122 headmasters of primary schools in Brunei Darussalam in order to assess their role in integrating ICT into the school curriculum. Two aspects of the findings were noticed. Although 79 % of the headmasters gave priority to the integration of ICT in the teaching and learning process, only 35 % of the headmasters allowed teachers an amount of time for appropriate use of ICT in the classroom. Another study in Brunei demonstrated different findings. Salleh and Kumar (2014) conducted a survey among 66 headmasters and 1138 teachers working in primary schools. The findings showed that only 47 % of school headmasters emphasized the importance of integrating ICT in their teaching lessons.

6 Conclusions and Implications

This study demonstrated the current situation of ICT regarding the education programs conducted in the member states of ASEAN and sought to synthesize the existing body of research on ICT in education among ASEAN member states. The study employed Van Dijk's causal and sequential model of digital technology as a conceptual framework to examine the digital divide in the integration of ICT in education among ASEAN member states regarding four aspects of access: motivation, material, skill and usage. Interestingly, the results of the research revealed that a digital divide, inequality of access to ICT in education, remains a challenge in all countries in ASEAN regardless of their levels of economic development and ICT development in education.

Implications for overcoming the digital divide for ICT in education in ASEAN

ASEAN member states have been facing problems of inequality in access to ICT in education in terms of motivation, material, digital skills, and usage access. The findings of this study have several policy and practical implications for various stakeholders in ASEAN [educational institution, local government, policy makers, the Association of Southeast Asian Nations (ASEAN)].

Implications for motivation access

Due to poor infrastructure and with no access to ICT, students living in rural areas of ASEAN member states report having computer anxiety symptoms. Schools should provide individual hands-on computer engagement activities in the classroom in order to create a feeling of familiarity and eventually reduce computer anxiety.

Implications for material access

In spite of the fact that Cambodia, Thailand, Indonesia, Malaysia, and the Philippines have either included ICT in their national education policy or have a stand-alone plan, there are still gaps at the material access stage. The problems include lack of infrastructure, low quality of computers and Internet connections. The government of these countries should allocate more resources to the out-of-reach areas, especially schools in rural areas, and support the investment of ICT infrastructures through the formulation of ICT national policies. Inequality of access to ICT materials still occurs in the transforming countries, where ICT has been fully integrated in the national education plan, for example in Singapore and Malaysia. Out-of-date hardware and software prevent teachers in Malaysia and Singapore from using ICT in their teaching. In order to ensure the comprehensive use of ICT in classrooms, schools should provide teachers with up-to-date ICT resources, including hardware and software.

Implications for Digital skill access

The teachers in almost all of the ASEAN member states reported a lack of ICT basic skills. Therefore, the Ministry of Education should provide continuous ICT training courses in order to increase the teachers' capacity to fully integrate ICT into the school curricula.

Implications for Usage Access

The findings showed that teachers reported a lack of time for integrating ICT infrastructure into the teaching process because of heavy workloads. Schools should provide organizational skill training programs in order to help teachers develop effective strategies for time management, and the planning and integration of ICT in the classroom. At the same time, schools should provide teachers with more time to utilize ICT in their teaching lessons. Moreover, there is only a small proportion of school principals that have high awareness of the integration of ICT in the school curriculum, hindering teachers from using ICT in the teaching and learning process. The Ministry of Education should create awareness and gain the commitment of school principals regarding the importance of the integration of ICT in the learning and teaching process. All in all, ASEAN should call on member states to report on their status of ICT in education and emphasize the importance of sharing ICT in education best practices and experiences within the ASEAN region.

Appendix 1: Diagram Showing the Study Selection Procedure

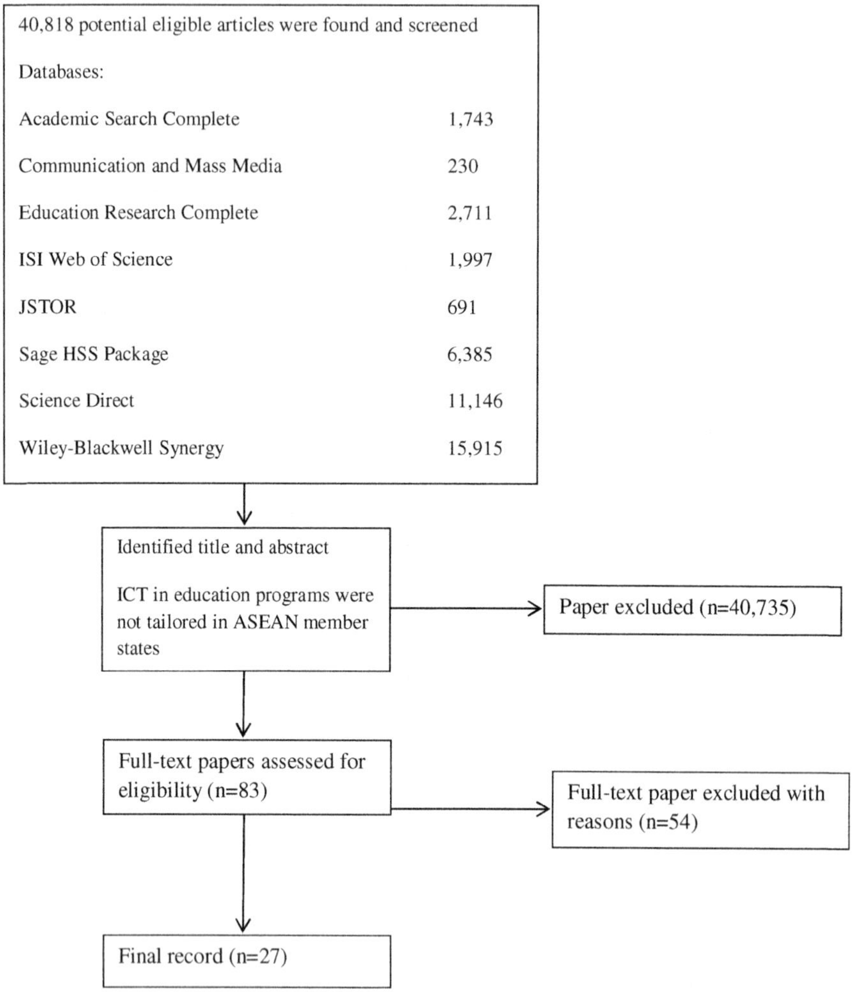

Appendix 2: Lists of Included Studies

No	Author/year	Country	Educational level	ICT tools	Methodology	Participant characteristics	Findings
1	Salleh and Kumar (2014)	Brunei Darussalam	Primary school level	ICT for learning programs	Quantitative	66 government primary schools 66 school principals, and 1138 teachers who were at that time working in government primary schools	47 % of school principles that placed emphasis on the integration of ICT in the teaching and learning process
2	Seyal (2012)	Brunei Darussalam	Primary school level	Computer, software, and Internet	Quantitative	122 Bruneian primary school principals	79 % of the principals prioritize the integration of computer technology in the classroom 69 % of the principals assist teachers in integrating ICT into the classroom 28 % of the principals offer teachers time to use ICT devices
3	Richardson et al. (2014)	Cambodia	Secondary school level	Computer and Internet	Quantitative	1137 students enrolled in three urban upper-secondary schools in Cambodia	Scarcity of computers, low accessibility, and computer anxiety
4	Dionys (2012)	Cambodia	NA	Desktop computers, DVD player, TV, projector, and netbook	Qualitative	Document	Teacher trainers have experienced computer anxiety

(continued)

(continued)

No	Author/year	Country	Educational level	ICT tools	Methodology	Participant characteristics	Findings
5	Shin et al. (2014)	Cambodia	Secondary level	Internet and computers	Quantitative	121 secondary school teachers in Cambodia	Limited ICT usage due to low ICT infrastructure and insufficient teacher training programs
6	Jayson (2011)	Cambodia	Preschool level Primary school level and secondary school level	Computer hardware and software, printers, scanners, digital cameras, digital recorders, and Internet	Mixed methods	379 Cambodian teachers were given the questionnaires and 17 Cambodian teacher trainers	Low ICT infrastructures and Internet connection
7	Rye (2009)	Indonesia	Higher education level	E-learning programs and Internet connection	Qualitative case study	Documents, observation and interviews with key informants including the students, the dean, the vice-dean, members of the research and development department, heads of the computer department and the regional office, the UT academic staff, and external lecturers and tutors of the Universitas Terbuka	Poor ICT infrastructure, resource management, and ICT plan in the rural areas of Indonesia
8	Safitry et al. (2015)	Indonesia	Not identified	Computer hardware, and software	Quantitative	30 elementary school teachers in Cipayung, Depok, West Java, Indonesia	Low competency level in using Microsoft Office, search engines, and e-mail Only 30 % of the teachers always use computers in the classroom

(continued)

(continued)

No	Author/year	Country	Educational level	ICT tools	Methodology	Participant characteristics	Findings
9	Copriady (2014)	Indonesia	Secondary school level	CD-ROM, computers, Internet, e-mail	Quantitative	874 Indonesian high school teachers	Motivation plays a crucial role in enhancing the use of ICT in the classroom.
10	Arokiasamy et al. (2015)	Malaysia	Secondary school level	Computers	Quantitative	520 secondary school principals in the state of Selangor and Wilayah Persekutuan, Malaysia	Low ICT skills including database use, spreadsheets, presentation/multimedia software, and the Internet
11	Aziz and Shah (2015)	Malaysia	Secondary school level	ICT learning tools	Qualitative	12 English teachers from an independent Chinese secondary school in Malaysia	Teachers have a heavy workload, lack of time. and low ICT skills Negative attitude toward using ICT in the classroom
12	Abdullah et al. (2009)	Malaysia	University level	Electronic distance learning and computer	Quantitative	500 students enrolled in undergraduate degrees at the University Technology Mara in Malaysia	Adult students experienced computer anxiety
13	Chan Yuen et al. (2011)	Malaysia	Undergraduate education level	Computer and Internet	Quantitative	70 pre-service teaching English as a second language (TESL) teachers in the Faculty of Education from a public university in Malaysia	Poor ICT facility and ICT technical service in schools
14	Hamzah et al. (2009)	Malaysia	Primary and secondary levels	Computer hardware and software and Internet	Qualitative	Seven focus group interviews were held in four schools Participants included ten teachers and seventeen students in two states, Selangor and Negeri Sembilan	An insufficient number of computers prevents teachers from integrating ICT in the learning and teaching process

(continued)

(continued)

No	Author/year	Country	Educational level	ICT tools	Methodology	Participant characteristics	Findings
15	Kaur and Sidhu (2010)	Malaysia	Undergraduate level	Computer, e-mail and Internet	Qualitative	30 part-time first-year students who enrolled in education course at the Faculty of Education in a private university located in the state of Selangor, Malaysia	Low computer and Internet accessibility
16	Yieng and Saat (2013)	Malaysia	Primary school and secondary school levels	Computer	Qualitative	Using data from Trends in International Mathematics and Science Study (TIMSS)	Low computer usage rate
17	Wan Zah Wan et al. (2009)	Malaysia	Secondary school level	Computers and Internet	Qualitative	Document analysis, interviews, and observations were conducted in three technology-rich Malaysian secondary schools	Teachers have heavy workload and lack of time to integrate ICT into the learning process
18	Gardner and Yaacob (2009)	Malaysia	Primary school level	CD-ROM	Qualitative	Observation and in-depth interviews with four or five teachers and students from four classes in four primary schools from two districts in Kedah Malaysia	Teachers report a lack of time to integrate the CD-ROM into English language lessons
19	Masagca and Londerio (2008)	The Philippines	Primary level	Computers, e-mail, and Internet	Qualitative	20 teachers who enrolled in the master's program of colleges and universities and employed by public and private schools in Luzon, the Philippines	Teachers with economic disadvantage and low digital skill have limited access to the Internet

(continued)

(continued)

No	Author/year	Country	Educational level	ICT tools	Methodology	Participant characteristics	Findings
20	Rodrigo (2005)	The Philippines	Primary and secondary school levels	Software, Internet, CD-ROM, and computer	Quantitative	79 public schools 74 private schools	High student-to-computer ratios Limited ICT infrastructure
21	Lim (2015)	Singapore	Secondary school level	ICT learning tools	Qualitative	21 teachers that teach subjects in areas such as language, humanities, mathematics, and the sciences	ICT is a time-consuming task Heavy workload prevents the integration of ICT in the classroom Teachers not only reported that ICT
22	Lim (2007)	Singapore	Primary school level Secondary school level Junior college level	ICT learning tools	Qualitative	Teachers and students in five primary schools, three secondary schools, and two junior colleges in Singapore	Teachers and students in Singapore place more emphasis on "improving examination results" or "making the grade" rather than integrating ICT in classroom
23	Choy et al. (2009)	Singapore	Higher education level	Computer hardware and software and Internet	Mixed methods	Questionnaires were distributed to 118 students enrolled in postgraduate teacher education programs in Singapore and in-depth interviews were conducted among 10 students	Out-of-date software and low-speed Internet
24	Pruet et al. (2014)	Thailand	Undergraduate and graduate levels	Tablet computer	Quantitative	213 grade two students in economically underprivileged regions of North Thailand	high learning competitiveness and high levels of anxiety toward tablet use among rural students

(continued)

(continued)

No	Author/year	Country	Educational level	ICT tools	Methodology	Participant characteristics	Findings
25	Tongkaw (2013)	Thailand	Undergraduate and graduate levels	ICT resources	Qualitative	Five stakeholders including higher education leaders, policy makers, technicians, lecturers, policy makers, and students that were at that time implementing ICT in local universities located in 41 provinces across Thailand	Low availability of ICT materials Poor management of ICT infrastructure in terms of plans and resource management Limited ICT technical support in classrooms
26	Akarawang et al. (2015)	Thailand	Primary school and secondary school levels	Computer hardware and software, and Internet	Mixed method	Questionnaires were distributed to 377 teachers from 35 schools in the northeastern part of Thailand In-depth interviews were conducted among 12 key informants, including 4 school directors and 8 teachers	Teachers reported a lack of the teachers' technical mastery of ICT skills in integrating ICT in the student learning process
27	Peeraer and Van Petegem (2011)	Vietnam	Preschool, and primary and secondary school level	Computers and Internet	Quantitative	783 teachers that were at that time working in five Vietnamese teacher education institutions	Lack of computer confidence and low basic computer, Internet, and maintenance and security skills

References

Abdullah, C., Ahmad, H., & Hashim, R. (2009). Attitudes toward ICT of electronic distance education (ePJJ) students at the Institute of Education Development, University Technology Mara. In J. Yang, A. Ginige, H. Mayr, & R.-D. Kutsche (Eds.), *Information systems: Modeling, development, and integration* (Vol. 20, pp. 222–228). Berlin, Heidelberg: Springer.

Akarawang, C., Kidrakran, P., & Nuangchalerm, P. (2015). Enhancing ICT competency for teachers in the Thailand basic education system. *International Education Studies, 8*(6), p1.

Anderson, J. (2010). *ICT transforming education: A regional guide*. Bangkok: UNESCO, Asia and Pacific Regional Bureau for Education. Retrieved March 24, 2011.

Arokiasamy, A. R. A., Abdullah, A. G. K. B., & Ismail, A. (2015). Correlation between cultural perceptions, leadership style and ICT usage by school principals in Malaysia. *Procedia—Social and Behavioral Sciences, 176*, 319–332. doi:10.1016/j.sbspro.2015.01.478

Aziz, A. B. A., & Shah, P. B. M. (2015). A case study: Obstructions encumbering the teacher's incorporation of ICT in English classrooms. *International Journal of Research—Granthaalayah, 3*(2), 57–68.

Baggaley, J., & Belawati, T. (2007). *Distance education technology in Asia*. Lahore: Virtual University of Pakistan.

Beeson, M., & Stubbs, R. (2012). Routledge handbook of Asian regionalism. London: Routledge.

Brownson, R. C., & Petitti, D. B. (1998). *Applied epidemiology: Theory to practice*. Oxford: Oxford University Press on Demand.

Cammaerts, B., & Van, A. (2003). Dominant digital divide discourses. In B. Cammaerts, Van Audenhove., G. Nulens, & C. Pauwels. (Eds.), *Beyond the digital divide: Reducing exclusion, fostering inclusion*. Brussels: Brussels University Press.

Chan Yuen, F., Sidhu, G. K., Shah, N. K. M., & Aziz, N. A. (2011). Pre-service teachers' training in information communication and technology for the ESL classrooms in Malaysia. *Turkish Online Journal of Distance Education (TOJDE), 12*(3), 97–108.

Choy, D., Wong, A. F. L., & Ping, G. (2009). Student teachers' intentions and actions on integrating technology into their classrooms during student teaching: A Singapore study. *Journal of Research on Technology in Education (International Society for Technology in Education), 42*(2), 175–195.

Clothey, R. (2012). *Post-secondary education and technology: A global perspective on opportunities and obstacles to development*. London: Palgrave Macmillan.

Copriady, J. (2014). Self-motivation as a mediator for teachers' readiness in applying ICT in teaching and learning. *Turkish Online Journal of Educational Technology, 13*(4), 115–123.

Cullen, R. (2003). The digital divide: A global and national call to action. *The Electronic Library, 21*(3), 247–257. doi:10.1108/02640470310480506

Dionys, D. (2012). Introduction of ICT and multimedia into Cambodia's teacher training centres. *Australasian Journal of Educational Technology, 28*(6), 1068–1073.

Dutta, S., Geiger, T., & Lanvin, B. (2015). *The global information technology report 2015*. Paper presented at the World Economic Forum.

Gardner, S., & Yaacob, A. (2009). CD-ROM multimodal affordances: Classroom interaction perspectives in the Malaysian English literacy hour. *Language and Education: An International Journal, 23*(5), 409–424. doi:10.1080/09500780802691751

Hamzah, I., Ismail, A., Tamuri, A. H., Embi, M. A., & Maimun, A. L. (2009). The emergence of new technology in Malaysian smart schools: Views of Islamic education teachers and students. *International Journal of Learning, 16*(4), 249–262.

iN2015 Steering Committee. (2015). Innovation, integration, internationalisation: Report by the iN2015 steering committee: iN2015 Steering Committee.

International Business Publications. (2012). *Singapore telecom industry business opportunities handbook*. Washington: International Business Publications.

International Business Publications. (2014). *Myanmar internet and e-commerce investment and business guide: volume 1 strategic, practical information and developments*. Washington: Global Investment Center.

ITU. (2015). Digital divide progress report: 15 year review. 2/9/2015, from https://itu4u.wordpress.com/2015/08/26/digital-divide-progress-report-15-year-review/

James, J. (2003). *Bridging the digital divide*. Massachusetts: Edward Elgar.

Jayson, W. R. (2011). Challenges of adopting the use of technology in less developed countries: The case of Cambodia. *Comparative Education Review, 55*(1), 008–029. doi:10.1086/656430

Kamel, S. (2010). *E-strategies for technological diffusion and adoption: National ICT approaches for socioeconomic development: national ICT approaches for socioeconomic development*. Hershey: IGI Global.

Kariyawasam, R. (2008). *International economic law and the digital divide: a new silk road?* Cheltenham: Edward Elgar Publishing.

Kaur, R., & Sidhu, G. K. (2010). Learner autonomy via asynchronous online interactions: A Malaysian perspective. *International Journal of Education and Development using Information and Communication Technology, 6*(3), 88–100.

Kinuthia, W., & Marshall, S. (2010). *Educational technology in practice: Research and practical case studies from the field*. Charlotte: IAP.

Lim, C. P. (2007). Effective integration of ICT in Singapore schools: Pedagogical and policy implications. *Educational Technology Research and Development, 55*(1), 83–116.

Lim, M. H. (2015). How Singapore teachers in a pioneer 'School of the Future' context 'deal with' the process of integrating information and communication technology into the school curriculum. *Australian Educational Researcher, 42*(1), 69–96. doi:10.1007/s13384-014-0153-0

Makaramani, R. (2013). *UNESCO-RCD at SSRU: ICE in education country report*. Malaysia: UNESCO Resource Training and Distribution Centre Regional Seminar.

Malaysian Government. (2012). Preliminary report: Malaysia education blueprint 2013–2025.

Masagca, J. T., & Londerio, N. M. (2008). Teachers' perspectives on the integration of information and communication technologies (ICT) in school counseling. *International Journal of Education and Development using Information and Communication Technology, 4*(4), 35–49.

Mondi, M., Woods, P., & Rafi, A. (2008). A 'uses and gratification expectancy model 'to predict students' 'perceived e-learning experience'. *Journal of Educational Technology and Society, 11*(2), 241–261.

Ness, D., & Lin, C.-L. (2015). *International education: an encyclopedia of contemporary issues and systems*. London: Routledge.

Norris, P. (2001). *Digital divide: Civic engagement, information poverty, and the Internet worldwide*. Cambridge: Cambridge University Press.

Organization for Economic Co-operation and Development. (2001). *Understanding the digital divide*. Paris: OECD.

Oxford Dictionaries. (Ed.) (2014).

Peeraer, J., & Van Petegem, P. (2011). ICT in teacher education in an emerging developing country: Vietnam's baseline situation at the start of 'The Year of ICT'. *Computers and Education, 56*(4), 974–982. doi:10.1016/j.compedu.2010.11.015

Peeraer, J., & Van Petegem, P. (2012). The limits of programmed professional development on integration of information and communication technology in education. *Australasian Journal of Educational Technology, 28*(6), 1039–1056.

Peña-López, I. (2015). OECD digital economy outlook 2015.

Pruet, P., Ang, C. S., & Farzin, D. (2014). Understanding tablet computer usage among primary school students in underdeveloped areas: Students' technology experience, learning styles and attitudes. *Computers in Human Behavior,*. doi:10.1016/j.chb.2014.09.063

Richardson, J., Nash, J., & Flora, K. (2014). Unsystematic technology adoption in Cambodia: Students' perceptions of computer and internet use. *International Journal of Education and Development using ICT, 10*(2), 63–76.

Rodrigo, M. M. T. (2005). Quantifying the divide: A comparison of ICT usage of schools in Metro Manila and IEA-surveyed countries. *International Journal of Educational Development, 25*(1), 53–68. doi:10.1016/j.ijedudev.2004.07.002

Rye, S. A. (2009). Negotiating the symbolic power of information and communication technologies (ICT): The spread of Internet-supported distance education. *Information Technology for Development, 15*(1), 17–31. doi:10.1002/itdj.20110

Safitry, T. S., Mantoro, T., Ayu, M. A., Mayumi, I., Dewanti, R., & Azmeela, S. (2015). Teachers' perspectives and practices in applying technology to enhance learning in the classroom. *International Journal of Emerging Technologies in Learning, 10*(3), 10–14. doi:10.3991/ijet.v10i3.4356

Salleh, S. M., & Kumar, L. (2014). Headmasters and information and communication technology: Approaches in making the connection. *Research and Practice in Technology Enhanced Learning, 9*(2), 349–362.

SEAMEO. (2010). *Status of ICT integration in education in Southeast Asian countries*. Bangkok: The Southeast Asian Ministers of Education Organization Secretariat.

Servaes, J. (Ed.). (2014). *Technological determinism and social change. Communication in a tech-mad world*. Lanham: Lexington Books.

Seyal, A. H. (2012). A preliminary study of school administrators' use of information and communication technologies: Bruneian perspective. *International Journal of Education and Development using Information and Communication Technology, 8*(1), 29–45.

Shin, H. S., Seng, S., & Choi, H. (2014). The role of teachers in enhancing information and communication technology-integrated education in Cambodia. *International Studies Review, 15*(2), 71–92.

Sorj, B. (2008). *Information societies and digital divides*. Corso Milano: Polimetrica S.a.s.

Symaco, L. P., & Brock, C. (2013). *Education in South-East Asia* (Vol. 20). London: A&C Black.

Tongkaw, A. (2013). Multi perspective integrations information and communication technologies (ICTs) in higher education in developing countries: Case study Thailand. *Procedia—Social and Behavioral Sciences, 93*, 1467–1472. doi:10.1016/j.sbspro.2013.10.065

UNESCO. (2013). *ICT in education, policy, infrastructure and ODA status in selected ASEAN countries*. Bangkok: UNESCO Asia regional Bureau for Education.

UNESCO Asia and Pacific Regional Bureau for Education. (2004). *Integrating ICT into education: A collective case study of six Asian countries*. Bangkok: UNESCO.

UNESCO Institute for Statistics. (2012). Computers-ISCED 1, 2 and 3: UNESCO Institute for Statistics.

UNESCO Institute for Statistics. (2014). *Information and communication technology in education in ASIA: A comparative analysis of ICT integration and e-readiness in schools across Asia*. Montreal: UNESCO.

van Deursen, A. J., & van Dijk, J. A. (2009). Improving digital skills for the use of online public information and services. *Government Information Quarterly, 26*(2), 333–340.

Van Dijk, J. A. (2005). *The deepening divide: Inequality in the information society*. Thousand Oaks: Sage Publications.

Van Dijk, J. A. (2006). Digital divide research, achievements and shortcomings. *Poetics, 34*(4–5), 221–235. doi:10.1016/j.poetic.2006.05.004

Van Dijk, J. A. (2009). One Europe, digitally divided. In A. Chadwick & P. N. Howard (Eds.), *Routledge handbook of internet politics* (pp. 288–304). London: Routledge Taylor & Francis Group.

Wan Zah Wan, A., Nor, H. M., Hamzah, A., & Alwi, H. (2009). The conditions and level of ICT integration in Malaysian Smart Schools. *International Journal of Education & Development using Information & Communication Technology, 5*(2), 1–7.

Wilson, E. J. (2004). *The information revolution and developing countries*. Cambridge: MIT Press.

Yieng, L. P., & Saat, R. M. (2013). Use of information communications technology (ICT) in Malaysian science teaching; A microanalysis of TIMSS 2011. *Procedia—Social and Behavioral Sciences, 103*, 1271–1278. doi:10.1016/j.sbspro.2013.10.456

A Socio-Cognitive Approach to the Communication of SME Support. An Exploratory Qualitative Study in Turkmenistan

Maya Velmuradova

Abstract Certain researchers call to reconsider *Communication for development and social change* as a problem of "Techniques and Society" (McArthur, Jouët, Bardini, Lohento, Kiyindou, Missé). Thus, the models of *social integration of innovations* are used here to study *how the new development assistance modules are accepted and appropriated by their final users in developing countries*. There is no need to prove anymore that users' reception and appropriation of the proposed service is critical to the development programs' effectiveness, be it in public health, environmental issues, small and medium enterprise (SME) development, or other. We synthesize the Anglo-Saxon and French models and distinguish the common determinant axes for the innovations or novelties reception: before their actual use (*acceptance* models) and after it (*appropriation* models). The common base appears to be the mental construction of the *sense of use*: the user mobilizes his representations "already there" (*anchoring* in existing representations, Moscovici, Jodelet, LeBlanc) and his *imaginary* (Flichy, Musso) to assess the associated *use benefits–costs*, the anticipated-perceived *use experience*. In the literature, this mental construction process is theorized as the formation of the *Perceived Value of Use* (*PVU*) (Jouet, Mallein and Toussaint, Mallein et coll., Toussaint, Boenisch, Assude et al, Nelson, Kim et al.). However, it is necessary to further explore this process. Eager to explore the PVU concept in detail, we conducted a two-phase field qualitative study within the *SME support* sector (UN Sustainable Development Goals 8.3 and 9.3), in Turkmenistan, Central Asia. Our results suggest that we can model the *PVU* formation mechanism, and its role in the *cognitive appropriation* and *acceptance* of the new support services by the final users. We propose to test the results of our research, within the context of other developing countries and of other sectors.

Keywords Communication for development and social change · Social utility services · Social integration of innovations · Cognitive appropriation · Acceptance · Perceived value of use

M. Velmuradova (✉)
Institute of Research in Information and Communication Sciences,
Aix Marseille Université, 13005 Marseille, France
e-mail: mayavelmuradova@gmail.com

J. Servaes (ed.), *Sustainable Development Goals in the Asian Context*,
Communication, Culture and Change in Asia 2,
DOI 10.1007/978-981-10-2815-1_7

1 Introduction

Communication for Development and Social Change (CDSC) is at the crossroads of public interest Communication and International Communication. The diversity of appellations and sectors makes this area even more complex and controversial. However, these different approaches and instruments have a single finality: the development of society. Missé and Kiyindou (2006) give the following definition of Communication for Development and Social Change:

> Public Communication makes part … of the field of general social change. … [The CDSC] … integrates some new actors: international organizations, businesses, NGOs … [and] … can be defined as a strategic mean for rationalizing the implementation of the social, economic and cultural development programs through the integration of the communication into their strategic planning.

Within this framework, the problem of our study is the "*need to understand the user*" (Kiyindou 2008). Indeed, it is well proven that the effectiveness of the development programs and messages depends heavily on the users' reception and appropriation of those. This, regardless of the sector: public health, environmental communication, economic development support, or others. The communicational strategy and its evaluation should take into account the local realities of the development program users, their contexts, and their perceptions vis-a-vis the development support messages and offers (Servaes et al. 2012). Why do users accept certain services and reject others? How do they integrate the new social utility services into their representations and their lives? and How do they appropriate those?

2 CDSC and the Social Integration of New Development Assistance Modules by Their Users

We will raise questions of Techniques, Society, and Uses. And what role for the Perceived Value of Use?

2.1 CDSC and the Social Integration of Novelties

The idea of using communication techniques to achieve development goals emerged almost simultaneously with the idea of supporting development, shortly after the Second World War, mainly in the USA. Research on propaganda, mass media, and their influence on public opinion underlined the potential ability of communication to change the existing social attitudes and lifestyles. These ideas quickly spread within the area of development support eager to resolve the problems of the post-war reconstruction on national and international levels.

First, CDSC has evolved under the influence of the dominant paradigm of the unilateral diffusionist "center—periphery" communication (Lerner, Schramm, Rogers), where "development" was defined as linear economic modernization (Spencer; Rostow). Then, from the late 1960s, the influence of critical theories—theories of dependency in Latin America, cultural imperialism (Schiller), critical pedagogy (Freire), and more recent critical approaches to the Information Society (Francophone approaches)—has given rise to the alternative paradigm of Participatory communication "periphery-center" and "periphery-periphery" (Quebral, Beltran; Daiz-Bordenave; Dagron; Mattelart; Miege; Missé; Kiyindou; Wolton; Badillo; Bourgeois; Proulx). This paradigm aims to take account of the structural factors (the empowerment approach, UNESCO) of the cultural and social diversity, as well as of the right to self-determination of local beneficiaries (the approach through participation and praxis, Freire). Today, a number of researchers defend the idea of a convergent CDSC (Servaes; Wilkins; Morris), where the *persuasive* communication mode (the public offer logic) and the *participatory* mode (the social demand logic) are seen as complementary rather than contradictory (for a review of approaches, see for example Gudykunst and Mody 2002; Lie and Servaes 2015; Missé 2004; Servaes 1999; Gumucio-Dagron and Tufte 2006; Wilkins 2008).

However, both of these communicational approaches to social change (persuasive and participatory) are more normative than explanatory. Thus, different techniques and strategies are prescribed here to make communication more efficient, such as interpersonal diffusion (Rogers 1983), edu-tainment (Singhal and Rogers 1999), engaging communication (Bernard and Joule 2004), communication-relation (Badillo et al. 2011), participatory communication and conscientization (Freire 1970), and others. As for the variables explaining the users' acceptance (or rejection) of a novelty, the above models primarily use the theories of attitude and change (Fishbein and Ajzen). The stages of decision making in change are described by Rogers (1983), Prochaska, and DiClemente (1983), Andreasen (1995) or Freire (1970); where the formed *attitude* (the combination of beliefs and their evaluation) is likely to influence the behavioral *intention* (Ajzen and Fishbein 1975; Rogers 1983; Andreasen 1995). However, the influence of the *attitude* on the final behavior is clearly not that strong (Rogers 1983; Snyder 2002; Bernard and Joule 2004). According to some statistics, only 3–15 % changes in behavior are observed, despite the formation of favorable attitudes toward the proposed development programs (Rogers 1983; Snyder 2002). This gap between the *perceptions* (knowledge, attitude), the *motivation* (interest), and the actual *behavior* shows the inadequacy of the existing theorization concerning the variables, the mental processes and the representations involved in the social integration (acceptance, appropriation) of the new support offered on the user side (Servaes 2013; Snyder 2002; Wilkins 2002, 2008). In this case, what explains the acceptance or the rejection of the new offers? The ideas, services, and programs for development and social change are most often used to transform the existing social reality. This process of adoption and integration can then be seen as a process of adoption, integration, and appropriation of innovations by the society. CDSC is indeed comparable to the "old question" of Techniques and Society (McArthur 1982; Jouët 1992; Bardini 1993; Lohento 2003; Rogers and Hart 2002; Kiyindou 2004, 2008; Missé and Kiyindou 2006), which offers some additional perspectives for analyzing this problem.

2.2 Techniques, Society, and Uses: The Social Integration of Innovations

Considering the parallel between the development and social change communication and the communication of innovations, the combination of Anglo-Saxon and European theories would help to better understand this communicational problematic (Bardini 1993). We adopt this broader perspective, in which the "communication of ICT" is seen as the communication of innovations in general (Bernard 2003). In this sense, the French theories of the communication of innovations have been influenced by sociological theories of uses. Here, the paradigm of "social autonomy" (De Certeau) replaced the technological determinism. However, more recently, the theories of uses have evolved under the influence of the sociology of innovations (Callon and Latours; Arkich; Flichy). Nowadays, it is the paradigm of the "socio-technical interaction" that forms a compromise between the technical and the social deterministic logics within the theories of uses (Jouët, Breton and Proulx; Miege) and the Communication of innovations (Badillo 2013).

Within this theoretical framework, we cross the different contributions emerging from the literature on uses, as applied to our problematic of the social integration of new techniques. First, there are three key moments in the socio-technical interaction: the *innovation conception* stage, its *social insertion* stage, and finally the *acculturation* of the innovation on a long term. The *use* being defined as the "utilization for specific purposes" (Breton and Proulx 2002), and the literature review shows the different ways of new uses construction at the social insertion stage. Within the practical (real life) uses construction, there is some proper "social logic," with its specificities: the slow and long-term temporality; the insertion into the pre-existing practices and systems; and the process of creative and pragmatic subjectivation of novelty by its user within the real use situation. This creative subjectivation process is defined as the "adaptation," the "domestication," the "re-invention," or the "appropriation" of innovation. This appropriation process appears as an iterative empirical-cognitive process: that is, a practical and a mental construction of the new uses by the users (Breton and Proulx 2002). The *cognitive appropriation* is the mental construction of the use and use experience, as they are represented or imagined by the user facing some new technique. The literature shows that this type of appropriation has certain modalities. First, there are different levels (or degrees) of appropriation: from *access*, *assimilation,* and *accommodation* to the *sociopolitical appropriation* (Toussaint 1992; Scardigli 1994; Breton and Proulx 2002; Proulx 2005, 2008). Furthermore, the representations of use and of its sense appear to be determinant by the social integration of innovations: both for the innovation adoption and for its effective use in situation (Perriault 1989; All Toussaint 1992; Mallein and Toussaint 1994; Chambat 1994; Jouët 2000; Breton and Proulx 2002; Millerand 2003). Finally, the cognitive appropriation appears to have some specific properties: the relation to the *identity*; the strong *interactivity* with the context, and the sociocultural environment (so called *situated cognition*); the *projectivity* of the sense of use (teleological and pragmatic sense, the object "instrument of"); and the symbolic charge influenced by the individual and the social *imaginary*.

Due to the particular historical development of the sociology of uses in France, the phase of the innovations *acceptance* has been unfairly dismissed from the socio-technical approach (whereas, the *appropriation* concept is principally linked to the effective use situation which comes *after* the innovation adoption). Yet a closer examination shows some strong relations between the *cognitive appropriation* and the *acceptance* of innovation. In particular, we could point out the mental construction of the *situated use project* (anticipated use), as described for example in the studies on the *techniques acceptability* (TAM, Davis et al. 1989). Here, the user attributes the sense of use, by anticipating in a projective and subjective manner the perceived *usefulness* and the *ease of use* in order to *accept* the new technology. Thus, the subjective and pragmatic mental construction of the new situated use takes place not only within the practical use in situ, but also during the *acceptance* phase: the use *anticipation*. Moreover, a number of authors already incorporate these evaluative judgments *before* the use (assessing the acceptability) and *within* the use (the cognitive appropriation by assigning the use meaning) into an integrated complex evaluative judgment (Swan and Clark 1992; Stewart 2003; Bobillier-Chaumon and Dubois 2009; Barcenilla and Bastien 2009; Mallein 1994, 1997, 2008, 2012).

That being said, three recurring pillars of such a judgment of use (anticipated or perceived) emerge from the literature: (1) the representations "already there" and anchored in the pre-existing representations and systems (Jodelet 1989); (2) imaging and imagination (Flichy 1995; Musso 2007); (3) the attribution of the subjective and pragmatic meaning to the situated use. Thus, both *acceptance* and *appropriation* of a novelty by its users pass by mental construction of the use meaning: where the user manipulates the perceived and the imagined to evaluate the use experience, its functional/symbolic benefits. This mental process of "constructing the subjective and teleological meaning of the situated use" appears in the literature as the formation of the Perceived Value of Use (PVU). Indeed, this concept appears in different works on the social integration of innovations, both within the approach on uses and their *appropriation*, as well as within the approach on innovation *acceptance* (Jouet 1987; Mallein et al. 1992, 1994, 1997, 2000, 2008, 2012; Hert 1998; Jurison 2002; Kaasinen 2005, Musso 2007; Kim et al. 2007; Boenisch 2009). The concept of PVU goes beyond the simplistic view of the perceived utility or of the rational trade-off: its conceptualization builds on a large body of work in consumer psychology and in consumer behavior. However, the formation of the PVU and its precise role in the social integration of innovations has not been studied in more detail. Our research question(s) are then as follows: *How is formed the Perceived Value of Use, on the users' side? What is the precise role of the PVU within the cognitive appropriation and the acceptance of the new social utility services by their users?*

2.3 Perceived Value of Use

The economic, psychological, and consumer behavior literature reveals a dispersed image and a transverse ambiguous concept of perceived value. We would first reinsert this concept in the historical context of the economic theory of value (Von Neumann and Morgenstern 1947, Debreu 1959), in which the concept of the exchange value has passed through three conceptual stages. The classic economists first see the exchange value as determined by the offer: it is the value-labor. The marginal economists seize the distinction posed by Aristotle between the *value of exchange* and the *value of use*. From this perspective, the *value of exchange* is subjective (perceived) and demand-driven, depending on the *value in use* for the consumer. Finally, the neoliberal economists put the exchange value as inter-subjective, e.g., formed during the interaction of supply and demand, through the strategies used by the market players. On the consumer side, the *Perceived value of exchange* (the price he is willing to pay, resulting from mental trade-off between the benefits and the costs of exchange) depends on his resources and preferences. The preferences are an expression of the *Subjective Expected Utility value* (SEU) that a consumer attributes to the product. This *Subjective Expected Utility value* (SEU theory, von Neumann and Morgenstern, Debreu) is seen here as the sum of the probable consequences of an action. The consequences (utilities) are projective and subjectively desirable (positive) or undesirable (negative) (Simon 1990a, b; Edwards and Fasolo 2001).

In cognitive, dynamic, and social psychology, the concept of *perceived value* evolves in different ways. Cognitive psychology builds on the *expected utility value* within the studies on *decision making* and on *problem solving*, aiming to investigate the limits of consumer rationality and simulate these cognitive processes (Simon, Tversky and Kahneman, Edwards). Dynamic and social psychology, however, studies the *perceived value* within the framework of *expectancy-value theories*, elaborated in the areas of *motivation* and of *attitude*. This is the *valence value* (Freud, Lewin, Allport, Peak Rosenberg, Fishbein). The *expected utility value* and the *expectancy value* are broadly defined in a similar way, that is as "the probability of a consequence x valence (positive or negative) of this consequence." Nevertheless, the cognitive concept of utility value has evolved under the influence of the positive paradigm (behaviorist and neo-behaviorist), where this value is considered as a magnitude or *intensity* ("to value") within the objective "mental mechanism" of valuation (Simon 1990b). On the other hand, dynamic and social psychology theories view the perceived value as the perceived "subjective *content*" and the "mental *structures*" ("a value"). Therefore, Gardial et al. (1994) distinguish between research on "the evaluation process indicators" and on the "evaluation outcomes."

However, consumer behavior works on the *perceived customer value* see this concept primarily as a "mental structure," as noted by Gutman (1981: 116). This is the case of the "globalists" models of *perceived value of exchange* (PVE) ("Trade-off benefits–costs," "worth," or "net value," Zeithaml 1988). It is also the case for the models of *perceived value of use*, PVU ("value of benefits," "consumption value," or "gross value"). More particularly, the PVU concept is studied as subjective *contents* within the "analytical" models, highlighting the multi-compositional benefits of use and their types (Sheth et al. 1991; Holbrook 1996, 1999; Filser et al. 2003; Aurier et al. 2004). It is also studied in terms of mental *structures* within the "hierarchical" models of use benefits. Here, the *attributes*, the *benefits,* and the *values* are connected within the hierarchical cognitive means-ends chains ("means-ends hierarchy" or "hierarchy of values," Woodruff and Gardial 1996; Woodruff 1997). This approach adopts the definition of the PVU composed by the *probability* of a desired consequence (*Instrumentality* of mean-benefit, its performance) and the *valence* of the desired end (*Importance* of the desired end value) (Gutman and Vinson 1979; Gardial and Woodruff 1996). The rare research approach on *perceived value of use* that explicitly explores its *intensity* (value as magnitude) is the *Desired Value Change Theory* (Flint et al. 1997, Flint and Woodruff 2001, Flint et al. 2002). Flint uses the "top-down" approach to the hierarchy of values (Gardial and Woodruff 1996) and defends the idea that within this means-ends hierarchy, the changes in consumer desired end states (end values) will lead to the changes in desired outcomes (benefits-means) and then in the desired attributes. These changes occur both in the *nature* and in the *intensity* of the desired value. The authors propose the *tension* as the main predictor of *change* in the *intensity* or the *nature* of the desired value (2002: 108–110). However, the question is how emerges this *intensity*, how the PVU goes from "0" to "medium" or "high"?

The *Instrumentality* variable being put apart, we thus identified in the literature four different meanings of the benefit desirability value that can be put on two axes.

Axe 1: Value as *content/structures* (benefits) *versus* Value as *intensity* (valence) (Gardial et al. 1994).
Axe 2: *Desirability* value (desirable benefits) *versus desired* value (desired benefits) (Tisdale 1961; Perugini and Bagozzi 2004).

Therefore, further exploration seems to be necessary, as to the *meaning* of the PVU for users, its *formation,* and its specific *role* in the social integration of services.

3 Formation of Perceived Value of Use and Its Role in the Social Integration of the New Development Support Services: Case of SME Support, EDP-USAID Project in Central Asia

Given the discussed limits of the existing theorization on the topic, the objective of our empirical study presented here was to understand "in depth" the formation of the PVU and its role in cognitive appropriation and acceptance of the new services by their users. The context studied here is the sector of SME support services: free advising and accompanying services, proposed within the USAID Enterprise Development Project (EDP-USAID) to the local SMEs in post-Soviet Central Asia (Turkmenistan).

3.1 Context and Methodology of the Study

The Turkmenistan is a country located in post-soviet Central Asia. This developing country that has gained its independence since 1991, has also the fourth world gas reserves, and has a complex geopolitical situation. Having obtained the status of positive neutrality (non-participation in military conflicts), this young country is focusing on the diversification of its industries and on its private sector development. Indeed, this latter was totally absent at the time of the USSR. While the SME support measures are adopted at the national level, some international organizations are also active within the private sector development through the various technical and financial assistance programs and initiatives. The Enterprise Development Project EDP has been the project funded and supported by the US Agency for International Development (USAID). Within the five countries of post-Soviet Central Asia, the project was implemented by the Pragma Corps (EU) via the 12 assistance centers with over 100 employees. The components of technical assistance to SMEs launched under the EDP project were the following: *Regional Trade Promotion* component; *Accounting Reform*—the application of international accounting standards; *Association Development*—professional associations networking; *Short Business Courses*; and the *Business Advising Services* (BAS) component—providing free consulting services and accompanying to small businesses. Our study focused on the new BAS component and the social integration (acceptance, appropriation) of its services by the local User SMEs in Turkmenistan. Indeed, the management consulting services being unfamiliar with local small businesses, the newly launched component had to cope with the low acceptance, and the low engagement of the user SMESs.

Thus, our research questions were as follows: In the case of these new SME support services, how the Perceived Value of Use is formed on the users' side?

What is the precise role of the PVU in their cognitive appropriation and their acceptance of the new service?

To answer these questions, we realized a qualitative multi-site case study, done by means of an iterative process of theoretical and empirical exploration. This "adductive approach" included four months of *participant observation* and *in-depth semi-structured interviews* with a selected sample of user SMEs. By adopting the perspective of a complex methodological individualism, we analyzed the cases of *valuation* of individual actors in their interaction with the organizational and surrounding environment. The first stage was the *participant observation* (4 months) by immersion, as a Recruiting Consultant and a Customer Relationship Manager (CRM). During this phase, we used the following sources: the database on user SMEs; field documentation; field notes of participant observation; open unstructured discussions with the local SMEs, and the support service providers concerning the support service acceptance situations. These qualitative data were analyzed later through lexical and thematic content analysis. We did also undertake an intervention/action research aimed to improve the user-SME acceptance and engagement into the new service, as part of our mission. In particular, we designed and implemented a perceived value-based engaging questionnaire. During the second phase, we conducted *in-depth interviews* and studied the selected sample of cases among user SMEs. The *semi-structured in-depth interviews* (50 min–3 h 40 min) were performed with user SMEs, including the open exploration of each "valuation-acceptance" situation of the service. After construction and testing of the interview guide, we explored the formation of the *Perceived Value of Use* by means of the *Value intensity mining technique*, which we developed for this occasion. This technique completes the existing *Laddering technique* (Reynolds and Gutman 1988), whereas the latter aims to explore the *hierarchy of value* in terms of its *contents* and the *means-ends structures*. Two different SME support providers were also interviewed in order to make a comparison. Narratives and other qualitative data were subject of the computer-assisted lexical and thematic content analysis intra-site and inter-site (Sphinx Lexica v. 5.1, NVivo v. 9.2).

For our multi-site case study, we used the selected sample of SME-users, in varying conditions (Yin 1984): EDP-BAS service users *versus* non-EDP users; the technical support services (business consulting and accompanying) *versus* the financial support (credits at preferential rates); not-for-benefit *versus* for-benefit services (commercial business consulting); rejection *versus* acceptance decisions; and narrative a posteriori *versus* in situ. The final sample included 9 organizational cases of user SMEs, resulting in 13 cases of valuation concerning 11 services (50 % —EDP-BAS services and 50 %—other SME support services).

The sample SMEs had the following characteristics:

Sector: 77 % services; 23 % agro; 15 % trade. Legal status: 54 % legal entity; 46 % auto-entrepreneur.
Company age: 73 % over 5 years.
Company size: 46 % TPE (<10 persons); 39 % PE (<50); 15 % ME (<200).

Respondents had the following characteristics:

Gender: 70 % men; 30 % women.
Age: 44.5 years on average.
Marital status: 69 % married, with children.
Education/experience: 77 % of higher education; with corresponding professional and entrepreneurial experience.
Position within the company: 77 % owners and 100 % higher level managers (decision makers or influencing role).

The cases of valuation included the acceptance and the rejection decisions:

the EDP-BAS service—2 cases "yes," 4 cases "no,"
other SME support services—6 cases "yes," 1 case "no."

We have distinguished four different scenarios of valuation-acceptance decisions:

"full acceptance,"
"rejection-to-acceptance,"
"acceptance-to-rejection," and
"full rejection."

3.2 *Results*

The aggregate results concern: (1) the *role* of the PVU in the social integration of the new service (acceptance and appropriation); (2) the *formation* of the PVU on the users' side; and (3) the *moderating factors* of valuation.

3.2.1 Role of Perceived Value of Use in the Social Integration of the New Service by the User SMEs

The PVU appears as the essence of the *cognitive appropriation*, anticipative, or in use. Indeed, the *subject*, the *object,* and the *project* of use are connected here within the mental construction of *meaning of use* (subjective, projective and situated). For a new object, it may be the imaginary *instrumentality of means* to achieve the existing *desired ends*.

There is a positive relationship between the formed *PVU* and the *acceptance decision*, where the latter is defined as the "intention to engage into the service." The *PVU* influences the *acceptance decision* via the mediator concept of the *perceived value of exchange* (PVE, the trade-off benefits–costs), which also involves the *perceived costs*. Indeed, there is a correspondence between the perception of the

PVU and the perception of *PVE* in 10 cases of valuation out of 13, and a correspondence between the perception of *PVE* and the *acceptance decision* in 13 cases out of 13.

3.2.2 Formation of the PVU at the SME-Users' Side

User SMEs show the three aspects of the *PVU*,—the *contents*, their *hierarchical structure,* and the *intensity* (result consistent with the *hierarchical theory of value*, Gardial and Woodruff 1996; Woodruff 1997). The *intensity* of PVU has a motivational basis: this is the value of the *desired*, rather than *desirable* benefits. This result supports as well the *hierarchical theory of value*. It is also consistent with the economic and the psychological motivational approaches of PVU, where the *valence* is measured as the "force of attraction of desired end state (goal)" (Jevons, Simon Edwards, Lewin, Peak, Rosenberg). The *desirability* appears then as the categorization of the object as potentially *desireable*: the question here is whether the end-state object is or not *able to be desired* (parameter: no/yes).

Thus, for every use benefit, the *PVU* appears to be formed by: (1) the *perceived Instrumentality* (qualitative correspondence and relative performance) of the benefit–mean to achieve the desired use goal and (2) the *valence* of this desired end (defined as the importance of desire/quasi-need). *Desires* ("quasi-needs") appear as "canalized expressions of needs" (Nuttin 1980). In our context, the principal use benefit of the advisory service designated by the SME-users is "external knowledge in business management." This benefit is related by *instrumental* cognitive links to the desired goals, such as some final states of the "performance," of the "status" or of the "self-realization" and others. These are subjectively *desired goals* that are "prioritized." Each has a more or less intense *valence* of attraction for each user. The *perceived value of use* of the "external knowledge in management" is then more or less intense (from zero to maximum), depending on: (1) perceived *Instrumentality* (does the "external knowledge in management" allow to reach the given *desired goal* and at what level, compared to other means such as "internal knowledge in management"); and (2) *valence* of that given desired goal (how much this goal is desired, its attracting force).

The results show that the *intensity* of this *valence of desired end* depends, in turn, on the motivational *tension* (result similar to Flint et al. 2002). This *tension* is perceived by the individuals as their *psychological distance* to *two* reference points: not only the *ideal point,* but also the *critical point* of affaires. *Valence of desired end* is "greater" when the individual perceives his current position as "far" from the *ideal point* (the recurring code "far from the desired goal," e.g., "far" from a certain level of the "enterprise performance," of the "status," or of the "self-realization."), and "close" to the *critical point* ("close" or "not far" from the "bankruptcy" or from the "failure" or from the "symbolic death"). And vice versa, the *valence of desired end* is perceived as "minimal" or "absent" when the individual feels subjectively to be “near” to the *ideal point* and “far” from the *critical point* of affaires. Thus, this result enriches the existing conceptualization of the motivational *tension*: we propose to add to the *distance* between the *actual* and the *ideal* states (dAI), the psychological distance between the *actual* and the *critical* states (dAC).

3.2.3 Less Moderating Factors

The moderating factors appear to be the *knowledge* (expertise, propensity for innovation), the *culture* (the cultural gap, the acculturation strategy), the *not-for-benefit finality*, and the *group dynamics*. In our context, these factors played a mixed role.

3.3 Model

All of these results are integrated in the model below:

This model aggregates all of our empirical propositions on the formation of the PVU and on its role in the social integration of new services. The model is based on qualitative data, following a grounded theory approach to the models of medium range (Fig. 1).

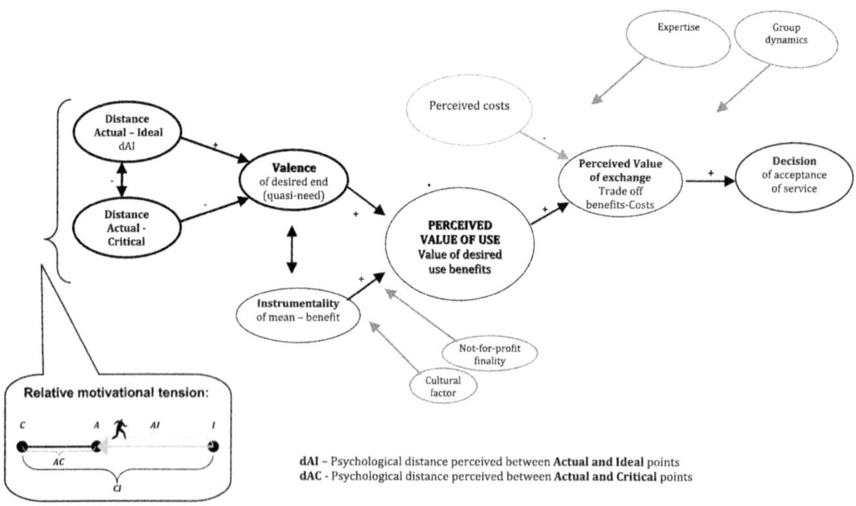

Fig. 1 Model of mental formation of the perceived value of use and its role in the social integration of new services

4 Conclusion

Communication for Development and Social Change (CDSC) is seen here as a question of techniques and society. The social integration of new social and development support services goes through the attribution of the "sense of use" by their end users, who mobilize the representations "already there" and the imaginary. This mental process of the subjective and projective assessment of the functional and symbolic benefits of use appears as the formation of the Perceived Value of Use. Our research focuses on the formation of the PVU and its role in the social integration of new services. The PVU appears as the subjective assessment of: (1) *Instrumentality* of benefit–mean (*correspondence*, *performance*) to reach the *desired end*; and (2) the *valence* of this desired end of use (quasi-need). The intensity of valence is based on the *motivational tension*, perceived by individuals as a *psychological distance* to two reference points: the "*Ideal*" point, but also the "*Critical*" point of affaires. The formed *PVU* constitutes the essence and the concrete expression of *cognitive appropriation*, connecting the *object*, the *subject,* and the *project* of use. Finally, there is a positive relation between the *PVU* and *acceptance* of new service; this latter defined as the *intention to engage* into the service. This relation passes through the mediator concept of *Perceived Value of Exchange* (PVE), defined as the *mental trade-off* between the *PVU (perceived Benefits)* and the *perceived Costs*.

The results presented here contribute to the communicational theories on the social integration of new technologies: the theories concerning in particular the *cognitive appropriation* (Breton and Proulx 2002; Millerand 2002, 2003), the *acceptance* and social roots of these (Mallein and Toussaint 1994; Mallein 2012; Leblanc 2007). They also provide insight into the reception of messages and services for social change in the developing countries of East and South (Missé; Kiyindou), but also in industrialized countries (influencing pro-social behaviors by engaging communication, Bernard and Joule). Moreover, these results enrich the understanding of the PVU concept in the theories of consumer behavior. They contribute in particular to the *Theory of hierarchy of values* (Woodruff and Gardial) and the *Theory of desired value change* (Flint), both theoretically but also methodologically: the *technique of the value intensity exploration*. Finally, these results can be used by practitioners of CDSC, managers, and practitioners of development and social marketing programs: as it concerns the analysis and measurement tools of engagement behavior of users, especially in situations of low valuation and demand of new services.

However, this study has also limitations: limited generalization of qualitative results; the qualitative and non-random sampling; the challenges of the "non-desirability" value measurement; the challenges of a multilinguistic environment. The next steps would be to perform tests and to apply these results in the context of other developing countries (for example, in Africa) and in other sectors. It would also be interesting to transpose these findings to the situations of "false valuation" and of the "impossible individual desirability," frequently encountered in the social/public utility service sectors such as in public health or the environment.

Acknowledgments The author thanks the USAID EDP (US) for technical assistance; the Foundation OSI–Global Supplementary Research Grant Program (US), and Schlumberger (France) for the partial financing via search donations; BGF-MAE Programme (France) and IRSIC-AMU Institute (France) for the technical and financial support. The author also thanks the anonymous reviewers for their comments and suggestions to improve this paper.

References

Andreasen, A. (1995). *Marketing Social Change.* San Francisco: Jossey-Bass Publisher.

Aurier, P., Evrard, Y., & N'goala, G. (2004). Comprendre Et Mesurer La Valeur Du Point De Vue Du Consommateur. *Recherche et applications en Marketing, 19*(3), 1–21.

Badillo, P.-Y. (2013). Les théories de l'innovation revisitées: une lecture communicationnelle et interdisciplinaire de l'innovation—du modèle 'émetteur' au modèle communicationnel. *Les Enjeux de l'information et de la communication, 14*, 19–34 [online] http://lesenjeux.u-grenoble3.fr/2013/Badillo/index.html

Badillo, P.-Y., Bourgeois, D., & Asdourian, B. (2011). Perspectives des nouveaux champs de la communication des organisations: Éléments à partir du projet européen Fire Paradox. *Les Cahiers du numérique, 6*(4), 167–180.

Barcenilla, J., & Bastien, J. M. C. (2009). L'acceptabilité des nouvelles technologies: Quelles relations avec l'ergonomie, l'utilisabilité et l'expérience utilisateur? *Le Travail Humain, 72*(4), 311–331.

Bardini, T. (1993). Diffusionisme, constructivisme et modèle technique: Ebauche d'une approche communicationnelle du changement technico-social. *Technologies de l'Information et Société, 5*(4), 367–391.

Bernard, F. (2003). La communication: enjeu de société, enjeu scientifique et enjeu de formation. Communication présentée au colloque: *Pour une refondation des enseignements de communication des organisations*, Paris, 25–28 Août.

Bernard, F. & Joule, R. (2004). Lien, sens et action: vers une communication engageante. *Communication et organisation, 24*, [online], http://communicationorganisation.revues.org/2918

Bobillier-Chaumon, M., & Dubois, M. (2009). L'adoption des technologies en situation professionnelle: Quelles articulations possibles entre acceptabilité et acceptation? *Le Travail Humain, 72*(4), 355–382.

Boenisch, G. (2009). Jacques Perriault, La logique de l'usage. Essai sur les machines à communiquer. *Questions de communication, 15* [online], http://questionsdecommunication.revues.org/1232

Breton, P. & Proulx, S. (2002). *L'explosion de la communication à l'aube du XXIe siècle* (2 éd.). Paris, la Découverte.

Chambat, P. (1994). Usages des TIC: évolution des problématiques. *Technologies de l'information et société, 6*(3), 249–269.

Davis, F. D., Bagozzi, R. P., & Warshaw, P. R. (1989). User acceptance of computer technology: A comparison of two theoretical models. *Management Science, 35*, 982–1003.

Debreu, G. (1959). Theory of Value: An Axiomatic Analysis of Economic Equilibrium, Yales University Press. Trad. Comar, J-M & Quintard, J. (2001). Théorie de la Valeur: analyse axiomatique de l'équilibre économique, Paris, Dunod.

Edwards, W., & Fasolo, B. (2001). Decision Technology. *Annual Review of Psychology, 52*(1), 581–606.
Filser, M., Plichon, V., & Anteblian-Lambrey, B. (2003). La valorisation de l'expérience en magasin: analyse de l'adaptabilité d'une échelle de mesure de la valeur perçue. In *Actes du 6ème Colloque Etienne Thil.*, CD-ROM, Université La Rochelle.
Fishbein, M., & Ajzen, I. (1975). *Belief, attitude, intention and behavior: An introduction to theory and research.* Reading, MA: Addison Wesley.
Flichy, P. (1995). *L'Innovation technique.* Paris, Editions La Découverte: Récents développements en sciences sociales Vers une nouvelle théorie de l'Innovation.
Flint, D. J., & Woodruff, R. B. (2001). The initiators of changes in customers' desired value. *Industrial Marketing Management, 30*(4), 321–337.
Flint, D. J., Woodruff, R. B., & Gardial, S. F. (1997). Customer value change in industrial marketing relationships: A call for new strategies and research. *Industrial Marketing Management, 26*(2), 163–175.
Flint, D. J., Woodruff, R. B., & Gardial, S. F. (2002). Exploring the phenomenon of customers' desired value change in a business-to-business context. *Journal of Marketing, 66*(4), 102–117.
Freire, P. (1997 [1970]). *Pedagogy of the Oppressed* (revised ed). New York: Continuum.
Gardial, S. F., Clemons, D. S., Woodruff, R. B., Schumann, D. W., & Burns, M. J. (1994). Comparing consumers' recall of prepurchase and postpurchase product evaluation experiences. *Journal of Consumer Research, 20*(4), 548–560.
Gudykunst, W., & Mody, B. (Eds.). (2002). *Handbook of International and Intercultural Communication* (2nd ed.). Thousand Oaks: Sage.
Gumucio-Dagron, A., & Tufte, T. (Eds.). (2006). *Communication for social change anthology: Historical and contemporary readings.* New Jersey: Communication For Social Change Consortium.
Gutman, J. (1981). A means-end chain model for facilitating analyses of product markets based on consumer judgement. *Advances in Consumer Research, 46*(2), 60–72.
Gutman, J., & Vinson, D. E. (1979). Value structures and consumer behaviour. *Advances in Consumer Research, 6*, 335–339.
Hert, P. (1998). *Jeux, écritures, espaces d'énonciations. Contribution à une étude anthropologique de l'usage d'Internet en milieu scientifique.* Thèse de doctorat en SIC. Strasbourg, Université Louis Pasteur.
Holbrook, M. B. (1996). Customer value—a framework for analysis and research. *Advances in Consumer Research, 23*, 1.
Holbrook, M. B. (Ed.). (1999). *Consumer value: A framework for analysis and research.* London: New York, Routledge.
Jodelet, D. (Ed.). (1989). *Les représentations sociales.* Paris: PUF.
Jouët, J. (1987). *L'écran apprivoisé: télématique et informatique à domicile.* Paris: CNET.
Jouët, J. (1992). *Pratiques de communication et changement social. HDR en SIC.* Grenoble, Univ. Grenoble 3.
Jouët, J. (2000). Retour critique sur la sociologie des usages. *Réseaux, 100*, 487–521.
Jurison, J. (2002). Perceived value and technology adoption across four end user groups. *Information Systems Evaluation Management*, 1.
Kaasinen, E. (2005). *User acceptance of mobile services—value, ease of use, trust and ease of adoption.* PhD thesis in Technology. Espoo: VTT Publications.
Kim, H., Chan, H., & Gupta, S. (2007). Value-based adoption of mobile internet: An empirical investigation. *Decision Support Systems, 43*(1), 111–126.
Kiyindou, A. (2004). Modèles d'intégration des pays du Sud dans la société de l'information: entre assistanat, insertion et intégration. In *Actes de Groupe de travail Sociologie de la communication. XVIIe congrès international des sociologues de langue française*, Tours, 5–9 juillet, 2004.
Kiyindou, A. (dir.). (2008). *Communication pour le développement. Analyse critique des dispositifs et pratiques professionnels au Congo.* Fernelmont: EME & InterCommunications.

Leblanc, C. (2007). *Représentations sociales et dispositifs socio-techniques: Nouveaux services, nouveaux usages*. Thèse de doctorat en SIC. Nice: Université de Nice—Sophia Antipolis.
Lie, R., & Servaes, J. (2015). Disciplines in the field of communication for development and social change. *Communication Theory, 25*(2), 244–258.
Lohento, K. (2003). *Usages des NTIC et Médiation des savoirs en milieu rural en Afrique. Mémoire de recherche DEA*, dir. J. Perriault. Paris: Université Paris.
Mallein, P. (1997). *La conception assistée par l'usage*. Rapport du conseil scientifique.
Mallein, P. (2008). Innover dans le sens d'usage. papier de travail. [online], http://www.tpr.asso.fr/Documents/Documents/cautic.pdf
Mallein, P. (2012). La conception d'innovation assistée par l'usage. (méthode CAUTIC). *Communication présentée au Séminaire G2I-INNOVACS*, Janvier, MSH Alples.
Mallein, P. & Toussaint, Y. (1994). L'intégration sociale des technologies d'information et de communication. Une sociologie des usages. *Technologies de l'information et société, 6(*4), 315–336.
Miège, B. (2008). Questionnement de la sociologie des usages comme voie privilégiée de l'approche des TIC. *Communication présentée au Congrès AISLF Sociologie de la Communication*. Istanbul, Juillet 2008.
Millerand, F. (2002). La dimension cognitive de l'appropriation des artefacts communicationnels (pp. 181–203). In Jauréguiberry, F. & Proulx, S.. Dir.. *Internet*: *nouvel espace citoyen*. Paris, Éditions l'Harmattan.
Millerand, F. (2003). *L'appropriation du courrier électronique en tant que technologie cognitive chez les enseignants chercheurs universitaires. Vers l'émergence d'une culture numérique?*. Thèse de Doctorat en Sciences de Communication. Montréal: Université de Montréal.
Missé, M. (2004). *Aporie de la communication sociale pour le développement*. Conférence-séminaire No 5. Grenoble, GRESEC, Université de Grenoble 3.
Missé, M. & Kiyindou, A. (2006). Introduction. In Missé, M. & Kiyindou, A. (Eds.) *Actes du 1er colloque international sur Communication et changement social en Afrique et dans les Caraïbes: Bilan et perspectives*. Douala, avril 2006, Grenoble, GRESEC, Université Stendal Grenoble 3, Université de Douala.
Musso, P. (2007). Partie I, Imaginaire et Innovation. In P. Musso, L. Ponthou, & E. Seulliet (Eds.), *Fabriquer le futur 2: L'imaginaire au service de l'innovation* (2nd ed.). Paris: Pearson Education France.
Nuttin, J. (1980). *Théorie de la Motivation humaine: Du besoin au projet d'action* (1st ed.). PUF: Paris.
Perriault, J. (1989). *La Logique de l'usage*. Paris: Flammarion.
Perugini, M., & Bagozzi, R. (2004). The distinction between desires and intentions. *European Journal of Social Psychology, 34*(1), 69–84.
Prochaska, J. O., & Diclemente, C. C. (1983). Stages and processes of self-change of smoking: Toward an integrative model of change. *Journal of Consulting and Clinical Psychology, 51*(3), 390–395.
Proulx, S. (2005). Penser les usages des technologies de l'information et de la communication aujourd'hui: enjeux—modèles—tendances. In L. Vieira & N. Pinède (Eds.), *Enjeux et usages des TIC: aspects sociaux et culturels* (Vol. 1, pp. 7–20). Bordeaux: Presses universitaires de Bordeaux.
Proulx, S. (2008). La sociologie des usages et après? In *Actes de Congrès AISLF Sociologie de la Communication*. Istanbul, Juillet 2008.
Reynolds, T., & Gutman, J. (1988). Laddering theory, method, analysis, and interpretation. *Journal of Advertising Research, 28*(1), 11–31.
Rogers, E. (1983). *Diffusion of innovations* (3rd ed.). Free Press: New York.
Rogers, E., & Hart, W. (2002). The histories of intercultural, international and development communication. In W. B. Gudykunst & B. Mody (Eds.), *Handbook of international and intercultural communication* (2nd ed.). Thousand Oaks: Sage.
Scardigli, V. (1994). Déterminisme technique et appropriation culturelle: évolution du regard porté sur les technologies de l'information. *Technologies de l'information et société, 6*(4), 299–314.

Servaes, J. (1999). *Communication for development. One World, multiple cultures*. Cresskill: Hampton Press.

Servaes, J. (Ed.). (2013). *Sustainability, participation and culture in communication. Theory and Praxis*. Bristol-Chicago: Intellect-University of Chicago Press.

Servaes, J., Polk, E., Shi, S., Reilly, D., & Yakupitijage, T. (2012). Towards a framework of sustainability indicators for 'communication for development and social change'projects. *International Communication Gazette, 74*(2), 99–123.

Sheth, J., Newman, B., & Gross, B. (1991). Why we buy what we buy: A theory of consumption values. *Journal of Business Research, 22*(2), 159–170.

Simon, H. (1990a). Invariants of human behaviour. *Annual Review of Psychology, 41*, 1.

Simon, H. (1990b). *Reasoning in human affaires*. Stanford: Stanford University Press.

Singhal, A., & Rogers, E. M. (1999). *Entertainment-education: A communication strategy for social change*. Mahwah, NJ: LEA Publishers.

Snyder, L. (2002). Development communication Campaigns. In W. B. Gudykunst & B. Mody (Eds.), *Handbook of international and intercultural communication* (2nd ed.). Thousand Oaks: Sage.

Stewart, J. (2003). The social consumption of information and communication technologies (ICTs): Insights from research on the appropriation and consumption of new ICTs in the domestic environment. *Cognition, Technology & Work, 5*(1), 4–14.

Swan, J. A., & Clark, P. (1992). Organizational decision-making in the appropriation of technological innovation: Cognitive and political dimensions. *The European Work and Organizational Psychologist, 2*(2), 103–127.

Tisdale, J. R. (1961). *Psychological value theory and research: 1930–1960*. PhD dissertation. Boston: Boston University.

Toussaint, Y. (1992). Historique des usages de la télématique. In P. Chambat (Ed.), *Communication et lien social* (pp. 193–212). Paris: Editions Descartes/La cité des Sciences et de l'Industrie.

Von Neumann, J. & Morgenstern O. (1947). *Theory of games and economic behavior*. Princenton.

Wilkins, K. (2002). International development communication : Proposing a research agenda for a new Era. In W. Gudykunst & B. Mody (Eds.), *Handbook of international and intercultural communication*. Thousand Oaks: Sage.

Wilkins, K. (2008). Development communication. In W. Donsbach (Ed.), *International Encyclopedia of communication* (pp. 1229–1238). New York: Blackwell.

Woodruff, R. B. (1997). Customer value: The next source for competitive advantage. *Journal of the Academy of Marketing Science, 25*(2), 139–153.

Woodruff, R. B., & Gardial, S. F. (1996). *Know your customer: New approaches to understanding customer value and satisfaction*. Blackwell Business: Cambridge, Mass.

Yin, R. K. (1984). *Case study research. Design and methods*. Newbury Park, CA: Sage.

Zeithaml, V. A. (1988). Consumer perceptions of price, quality, and value: A means-end model and synthesis of evidence. *Journal of Marketing, 52*(3), 2–22.

Structural Reform Empowers Sustainable Development and Eradicates Poverty

G. Arockiasamy, Mallika Vijayakumar, Sujeevan Kumar and J.B. Anna Asheervadham Mary

Abstract Many countries in the world are projected in the global media to enjoy the benefits of globalization and advanced technology as the blessing of high economic prosperity. In recent years, the real picture emerged when economic crisis even afflicted the well-developed Western countries. It remained as a pre-warning to many developing counties in the world to assess their developmental strategies and economical pursuits. The recent collapse of the economy of Greece is a global lesson for many countries. Some countries blamed their governments adopting a capitalistic market-driven Western model. Others blamed the peoples' inabilities or lack of technological adaptation as the cause for poverty. The concept of poverty is related to development. Social development is broadly understood as a process where people are involved to raise their standard of living by means of activities designed by their free will. So far almost all governments in the world proclaimed this type of development as their agenda and coveted the political powers. How far they have succeeded in providing this sustainable development and empowerment? To understand the situation, we need a clear understanding of the concept of empowerment. This chapter will explore the meaning of empowerment and its intricacies from its ideological origin from Brazilian educationist Paulo Freire, who defined that a people's empowerment can be achieved only through peoples' awareness and participation. Our own renowned economist and Noble Prize winner Amartya Sen says that poverty is not simply the deprivation of basic material needs but concerns significant development in all dimensions of people's life. A quantitative research study is designed to assess the status of empowerment and progress in the present life through conducting a survey of sustainable development. The communication strategy and the structural change needed are also assessed in this research method. The findings of this study will provide us with a policy-related theoretical base. This will provide some information based on which this research can make recommendations and propose a structural reform for achieving sustainable development. Lastly, this chapter will present a method how to eradicate poverty by harnessing the political governance system through

G. Arockiasamy (✉) · M. Vijayakumar · S. Kumar · J.B. Anna Asheervadham Mary
Karunya University, Coimbatore, India
e-mail: arockiasamy@karunya.edu

J. Servaes (ed.), *Sustainable Development Goals in the Asian Context*,
Communication, Culture and Change in Asia 2,
DOI 10.1007/978-981-10-2815-1_8

reforming existing structure. It can provide proper adoption of technology and media as a participatory means in the hands of local people.

Keywords Empowerment · Development · Social reform and poverty · Communication

1 Introduction

The present time is considered to be an advanced digital age because we have crossed the heavy inoculate effects of media age after industrialization and modernization era. We are living a life dominated by "information." The recent technological influence has penetrated every society and intends to push them forward to form an information society. No doubt we are living in the age of infotainment. Hamelink (1998) notes that "in a zero sum society new values will evolve, new social relations will develop." All these developments are integral to both developed and developing countries. At the same time, all these countries faced economic crisis in the recent past, which have become an area of great concern. The fundamental question is about the reality of social development. In spite of technological progress, no society has reached a sustainable development (Servaes 2014). It is vital to examine the basic concepts.

Social development is broadly understood as a process in which local people participate in activities designed by their own choice with clear objectives to raise their standard of living by breaking the clutches of oppressive forces. The people find solution to their own sociocultural and political problems. They themselves direct their community to a happy life in all aspects. The above understanding leads to several assumptions: All are bestowed with equal power to decide their own development goals and strategies. They have the rights to enjoy the fruits of such efforts. So far almost all governments in the world proclaim this type of development as their agenda and thus covet the political power. How far have they succeeded in providing this sustainable development?

2 Literature Review

The India country report 2013 of SAARC development goals records in terms of two indicators. The first one is the income index recorded as Gini coefficient, which was 36.8 for India in 2010–2011. The second indicator is the quintile income ratio. For India, it is 5.6, which shows the inequality between the top and the grassroots, and is far lower than in many other Asian countries. Segar's (2013) study on the concept of women empowerment among female students of higher education in Madras Christian College, Chennai, found that factors such as freedom to make decisions, education, employment, and opportunity for political participation can facilitate empowerment of women.

A study on the mental health status of rural adolescent students of Erode district, by Venkatesh and Prabaker (2014), found that there was a lack of access to health facilities and opportunities among both men and women, thus leading to poor mental health among rural adolescents. Another study studied the linkage between work, family conflict, and women empowerment (Smitha 2013). It is noted that most of the women with non-traditional gender role attitudes experienced family conflict with high intensity.

The recent Noble Prize winner Angus Deaton from Princeton University has researched on poverty and welfare, and his findings reveal the tremendous health problems among adults and children in India, in spite of progress. He noted that half of the children in the country are still malnourished and many people in the world are affected by poverty. Most of the above studies make us wonder whether the prediction of Malthus, which states that growth of poverty and social conflict, is inevitable. These findings need to be tested once again for the present generation.

3 Objectives of the Study

Media and governmental agencies project that developing countries are fast growing in the economic and welfare index. But some critics claim that this is not the case. This study attempts to find out the existence of this disparity and inequality, as well as whether it hinders sustainable development. A number of factors that interplay in the process must be studied.

The future of the country's development is in the hands of the younger generation, and therefore, this study aims to assess the existence of inequality among students from city as well as rural environments. It can shed light on the reasons for the failure of *sustainable development*. Another objective is to propose a reform to tackle the problems faced by developing countries.

4 Conceptual Framework

See Fig. 1.

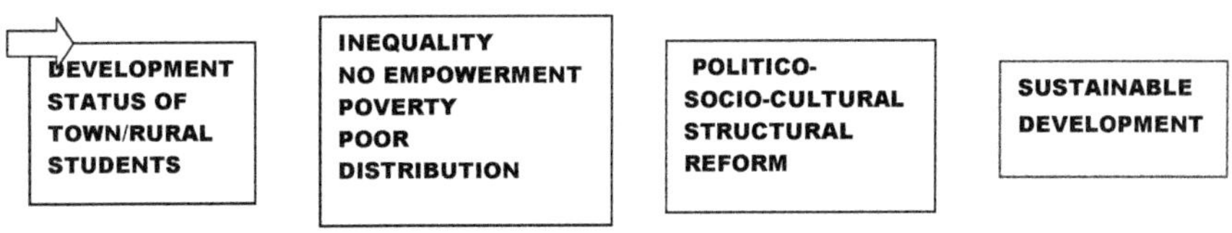

Fig. 1 Conceptual framework

5 Methodology

In order to study the presence of inequality and disparity in different economic levels in the present generation, the student community of both urban and rural areas was selected as the sample. A convenient sampling method was used in the respective areas, and a survey was conducted among both urban and rural students. The study also provided space to record their opinions and reasons for ongoing development activities and resource utilization by both private and government agencies. A comparative and statistical analysis was carried out to trace the causes of failure and thus to recommend a solution to rectify the problem.

6 Findings and Discussion

This study covers respondents from both urban and rural areas. The majority of the respondents were male (58.3 %) versus 41.7 % female respondents. Out of the total respondents, 63.3 % come from a rural background, while 36.7 % grew up in an urban setting. Age-wise, 83.3 % were between 18 and 25 years of age, while 16.7 % come from the group of above 26 years of age. 68.3 % of the respondents earn less than Rs.100,000 a year, while 31.7 % of the respondents earn more than Rs.100,001 (100 Indian Rupee = @1.5$) (Table 1).

The percentage of respondents who can afford to eat out at least three times a week amounts to 70 %, while those who eat out more times is only 30 %. The percentage of respondents who had private schooling is 61.7 %, 37.3 % studied in government schools and 5 % of the respondents studied in an international school,

Table 1 Frequency and percentage of some socio-demographic variables

Value	Label	Frequency	Percentage
Gender	Male	35	58.3
	Female	25	41.7
Place	Rural	38	63.3
	Metro	21	36.7
Age	18–25	50	83.3
	26–29	10	16.7
Income	Less than 100,000Rs	41	68.4
	100,000 and above	19	31.7
Eating out in a week	3 times	50	70
	4 times	8.3	8.3
Education schooling	Private school	37	61.7
	Abroad	1	1.7
College	University private colleges	31	52.3
	Abroad	1	1.7

Table 2 Frequency and percentage of variables on government versus private issues

Government infrastructure	Police surveillance	24	40
	Electricity	6	10
Government development	Strongly agree	3	5
	Agree	28	46.7
Private development	Bank	44	73.3
	Private finance corporation	2	3.3
View on government infrastructure	Developing	37	61.7
	Developed	6	10
View on private infrastructure	Highly developed	7	11.7
	Developing	31	51.7
Private efficiency	Satisfying	26	43.3
	Highly satisfied	4	6.7
Government efficiency	Highly satisfied	4	6.7
	Satisfied	29	48.3
Web information available	Google	50	83.3
	Blog	2	3.3

while 1.7 % studied in schools outside of India. The percentage of respondents who studied in government colleges is 26.7 %, while 52.3 % studied in private colleges (Table 2).

In terms of government infrastructure within 20 km around their locality, the rural population enjoys 50 %, while the urban population enjoys 40 % of the governmental infrastructure. 51.7 % of the respondents agreed that the government infrastructure development is good, while 49.3 % disagreed. In terms of private infrastructure up to 20 km around their locality, the rural population enjoys 18.4 %, while the urban population enjoys 81.6 %. 18.3 % feels that the status of the government infrastructure is developed, while 81.7 % feels the opposite. 28.4 % of the respondents are satisfied with the services provided by private companies, while 71.6 % are dissatisfied (Table 3).

According to the respondents, 55 % are satisfied with the services provided by the government, while 45 % were dissatisfied. The percentage of respondents who

Table 3 Frequency and percentage of technology, entertainment, and government variables

Technology available	Cell phone	29	48.3
	None of them	3	5
Development impact	Agree	28	46.7
	Strongly disagree	2	3.3
Entertainment available	Playing	13	21.7
	Playing instruments	3	5
Reasons for poor development	Poor government	38	63.3
	Lack of initiative	10	16.7

turn to the Internet for infotainment is 91.7 %, while 8.3 % do not. The percentage of respondents who incorporate technology in their daily activities is 95 %, while 5 % of the respondents do not use any form of technology. 65 % of respondents agree that the impact of technology is spoiling young minds, while 35 % of respondents disagree. 63.3 % of respondents say that the present status of development is due to poor government facilities, and 16.3 % feel that it is due to a lack of initiative. 11.7 % feel that there is no opportunity provided, while 8.3 % say that it is due to the wrong business. These descriptive data indicate that some development of infrastructure is provided by government and private efforts, but overall the feeling remains that there is inequality in the development sector and that rural people are neglected.

6.1 Chi-Square Test

There is a high significance in the variables studied under the development scheme, such as government infrastructure, private infrastructure in their area, status of government infrastructure, type of entertainment, and the reasons for the social condition in which the respondents live. There is a significant relationship in the variables such as development of private infrastructure and efficiency of private and government companies. The impact of technology in spoiling the minds of young people is also significant. Development is visible, but who benefits from the development is determined by the distribution of wealth. This unequal reality is expressed in the findings (Table 4).

6.2 Correlation Analysis

The findings of the Pearson's correlation test are presented in Tables 5 and 6. Income and place are having a positive correlation of 0.603. Also the place of living has a positive correlation with private companies as 0.365. Government infrastructure and government companies are significantly correlated as 0.426, with respect to the efficiency of private companies as correlated with government companies that results in 0.439.

The impact of technology on students is significantly correlated with their college education 0.314 and correlated with social development condition 0.343. Entertainment is also correlated with private companies' efforts shown as 0.355. Thus, this significance proves that there is development thanks to the efforts of government and private companies. Education plays a vital role in development. The development in rural areas is minimal, and such condition is due to poor government facilities and lack of initiative and no opportunity, and therefore, the present system will not help in reducing poverty and will fail to boost sustainable development. A structural reform is the only solution to solve this problem (Fig. 2).

Table 4 Chi-square significance of different variables

Variables	Chi-square	df
Gender of the students	1.667	1
Place of the students	13.333	3
Age of the students	65.333*	3
Income of the students	8.4	3
Eating out by the students	22.533*	3
School education of the students	56*	3
College studied by the students	30.8*	3
Government infrastructure in their place	12	3
Development in their place	38.533*	3
Private infrastructure available	76.4*	3
View on government infrastructure	44.933*	3
View on private infrastructure	24.667*	3
Private infrastructure efficiency	16.66*	3
Government infrastructure efficiency	22.267*	3
Web information used	109.2*	3
Technology usage	29.333*	3
Impact of technology on young minds	24.667*	3
Entertainment experience of students	32.533*	3
Reasons for society condition in their area	47.867*	3

* Chi square for 1 df (at 0.001 level) = 10.8 and 3 df = 16.3

Table 5 Correlation significance of variables

Variables	Income per annum		Government companies		Private companies	
Place of living	Pearson's correlation	0.603**	Pearson's correlation	0.296	Pearson's correlation	0.365**
	Sum of square and cross product	35,600	Sum of square and cross product	15,600	Sum of square and cross product	22,667
	N	60	N	60	N	60
Government infrastructure	Pearson's correlation	0.131	Pearson's correlation	0.426**	Pearson's correlation	0.159
	Sum of square and cross product	7500	Sum of square and cross product	21,750	Sum of square and cross product	9583
	N	60	N	60	N	60
Government companies	Pearson's correlation	0.081	Pearson's correlation	0.000	Pearson's correlation	0.439**
	Sum of square and cross product	4900	Sum of square and cross product	0000	Sum of square and cross product	27,750
	N	60	N	60	N	60

** Correlation is significant at the 0.01 level (1-tailed)

Table 6 Correlation significance of variables

Variables	Types of college education		Types of entertainment	
Impact of technology on students	Pearson's correlation	0.314**	Pearson's correlation	0.011
	Sum of square and cross product	11,800	Sum of square and cross product	400
	N	60	N	60
Social development condition	Pearson's correlation	0.343**	Pearson's correlation	0.066
	Sum of square and cross product	14,300	Sum of square and cross product	2600
	N	60	N	60
Private companies development	Pearson's correlation	0.308**	Pearson's correlation	0.355**
	Sum of square and cross product	16,750	Sum of square and cross product	13,933
	N	60	N	60

** Correlation is significant at the 0.01 level (1-tailed)

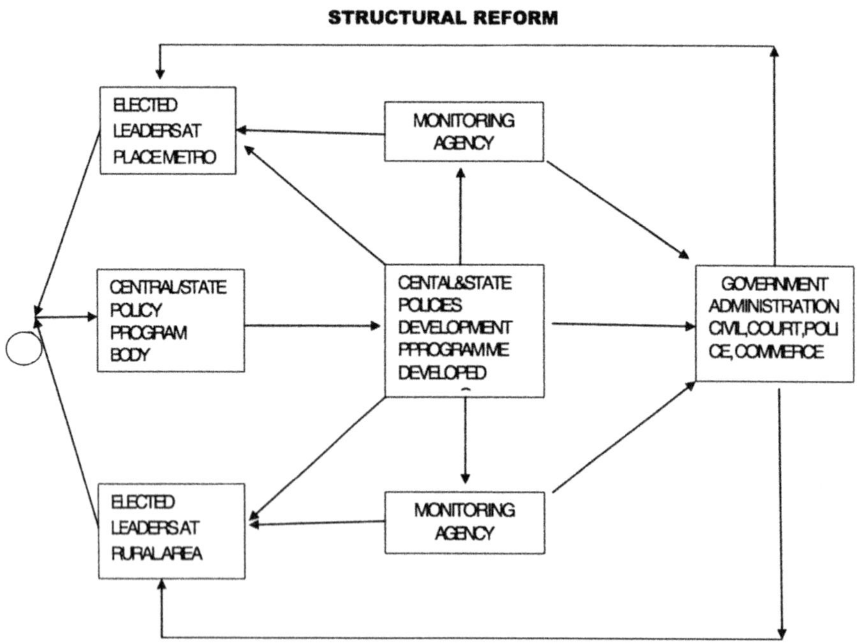

Fig. 2 Structural reform

In order to remove disparities and ensure a balanced distribution and empowerment of people for sustainable development, some of the existing evils in the system have to be removed. They are imbalanced development between rich countries and poor countries globally and similarly within the country. A balanced development between cities and local remote areas must be ensured. The nexus between rich elite business people and political leaders with the help of the administrative setup breeds corruption and tends to expand the gap between the haves and have not. This nexus must be broken in the new reform efforts. The leaders promise one thing and do another thing and maintain the status quo so that the disparity remains in all areas of development. This study also reaffirms the poor development and imbalance in distribution on several counts. Therefore, a new structural reform is crucial.

7 Conclusion

Empowerment is based on a power shift from political parties to local peoples. Therefore, a committee elected by the people should frame the policy and development according to their needs and importance to enhance an equal distribution. These development efforts will work for the total development of the periphery area of every state as well as the whole nation. Government, bureaucracy, media, and police should work as mediators in implementing these plans. A monitoring agency needs to be established to check the distribution of power and wealth in a balanced way. Such a development will enhance empowerment and in turn reduce poverty. This reform can induce major opportunities for both rural and urban dwellers to appropriate change where need arises. It also includes a number of measures to reduce the accumulation of wealth and to strengthen the distribution system. This can be tried as a pilot project in a specific area first and later broadened to assess sustainable development. That way the universal objective of humanity is geared toward the enrichment and welfare of all human life.

References

Deaton, A. (2015). *Angus Deaton, Nobel prize in Economics*. The Times of India, Coimbatore, October 1, 2015.

Hamelink, C. (1998). *ICT's and social development: The global policy context*. Paper presented at UNRISD conference on Information technologies and Social Development, Geneva.

Servaes, J. (Ed.). (2014). *Technological determinism and social change. Communication in a tech-mad world*. Lanham: Lexington Books.

Segar (2013). *The concept of women empowerment among female students of higher education, rural empowerment for sustainable development*. Chennai: Hikey Media.

Venkatesh Babu & Prabaker, S. (2014). A study on mental health in relation with personality of adolescent higher secondary school students, *Golden Research Thoughts*, http://aygrt.isrj.org.

Yada, Nisha (in press). "Globalization and structures of aspiration a psycho-cultural perception of indian middle class studies" in *Sociology, Journal of Department of Sociology*, University of Rajasthan, Jaipur, India.

http://www.marist.edu/admission/graduate/information-systems/technologies-influence-business.html
http://businesscasestudies.co.uk/anglo-american/new-technology-development-in-the-primary-sector/influences-on-process-development.html#axzz3iiT5ldXm
http://www.sky-bolt.com/TechnologyAndTraining.htm
https://books.google.co.in/books?id=SLDyTYc0r9oC&pg=PA144&lpg=PA144&dq=technology+used+in+business+influence+and+development&source=bl&ots=hreb-t8cZv&sig=4rAXuqN4uvMgJ_qiZga4g1SQGwQ&hl=en&sa=X&ved=0CFAQ6AEwB2oVChMIvuzLsKqoxwIVCB-OCh2khgHz#v=onepage&q=technology%20used%20in%20business%20influence%20and%20development&f=false
http://www.balanceinteractive.com/blog/influence-technology-culture
http://www.huffingtonpost.com/2013/09/09/gmc-the-future-is-now-_n_3843405.html?ir=India&adsSiteOverride=in
http://www.pcworld.com/article/152683/tech.html
http://matadornetwork.com/life/24-incredible-new-technologies-youll-see-by-2021/
http://www.popsci.com/
http://www.digitaltrends.com/cool-tech/
http://www.jemwindowcleaning.com/the-industries-newest-technology-is-available-at-jem-window-cleaning-today/

Conclusion: Are the SDGs "Sustainable"?

Jan Servaes

Abstract This concluding chapter argues that a successful implementation of the SDGs will have to address at least four challenges: (a) How can we bring together the right stakeholders at the right time in the right place?, (b) How do we make difficult trade-offs?, (c) How do we build in accountability and transparency for action?, and (d) How to organize this in a participatory and democratic way? It argues that a participatory approach is about dialogue, participation and the sharing of knowledge and information in a specific context, and briefly positions seven new challenges which emerged during the last decade, as a consequence of globalization, media liberalization, mediatization, rapid economic and social change, and the emergence of new social media and information and communication technologies (ICTs). It reiterates that, in the final analysis, the SDGs appeal to moral and ethical principles which we should share as people on this planet.

Keywords SDGs · Sustainability · Participation · Strategies

> How much healthier all our societies would be if they were based on value systems that truly advocated sustainability rather than unlimited growth! A society where people help each other out in hard times, a society where power is shared rather than fought over, that reveres and respects nature rather than controlling and using it as a resource, a society unsullied by the poisons of craving (*tanha*), and a society steeped in spirituality and wisdom (Sivaraksa 2009: 278).

The transition from MDGs to SDGs is ongoing. It remains work in progress. As one of the architects of the MDGs, Jan Vandemoortele (2011: 15), who was co-chair of the UN inter-agency group that put the MDGs together, contends: "Any revision of the MDGs will face pitfalls and challenges. … The new set (of SDGs, JS) should aim to include fewer targets than today, focused exclusively on ends for which the concept is clear, the indicators are solid and the data are robust. It should project ambitious yet feasible targets. The set should not be turned into aspirational

J. Servaes (✉)
Bangkok 10110, Thailand
e-mail: 9cssc9@gmail.com

J. Servaes (ed.), *Sustainable Development Goals in the Asian Context*,
Communication, Culture and Change in Asia 2,
DOI 10.1007/978-981-10-2815-1_9

goals or transformed into a normative statement." That statement was obviously made before the new set of SDGs was negotiated and announced, because today we know that the SDGs contain more targets, some of which are as unclear as the previous MDGs (UN 2014). So it remains to be seen whether the SDGs will be more successful.

In addition, the Dutch diplomat Van Norren (2012: 830) pleads for a focus on action or strategy instead of goals: "A post-2015 system should not be formulated in terms of goals but rather in terms of actions or processes. The result-based management method of the MDGs entails the danger of quantity before quality and an 'ends justifies the means' approach". She argues for gradual change rather than tabula rasa: "The indicators of the MDGs still have value and could be used in a different way (to measure progress and inequality, for example). Missing themes can be incorporated; clusters can be reshaped; interconnectedness can be made more visible; and the goal orientation can be attenuated by a strategy orientation" (Van Norren 2012: 834).

In the light of the 2030, Agenda for Sustainable Development, and in order to achieve the goals set within, the global community should no longer participate in a system that highlights division and inequality. The UN Secretary-General's High-level Panel on Global Sustainability (2012), and related working documents (SDSN 2015; UNSC 2015), note that all cultures contribute to sustainable development, which is in itself a reflection of the progress made since the implementation of the MDGs. For example, indigenous peoples have a much higher profile in the SDG outcome document—they are directly mentioned several times within the text —more than they did in the MDGs. Non-discrimination and non-exclusion are cornerstones to the successful application of the 2030 Sustainable Development Agenda. However, "what cannot be resolved by the post-2015 system, nor by any other model, is the reshaping of global power relations" (Van Norren 2012: 834).

A piecemeal approach in favor of gradual change also implies a realistic multi-stakeholder strategy, involving government, the private sector, and civil society. Multi-stakeholder partnerships are the "glue" that will hold this process together and will be the only way of ensuring these ambitious goals are met. A successful implementation of the SDGs will therefore have to address at least four challenges:

- First, how can we bring together the right stakeholders at the right time in the right place?
- Second, how do we make difficult trade-offs?
- Third, how do we build in accountability and transparency for action?
- Fourth, how to organize this in a participatory and democratic way?

1 CSSC and Participation

In line with this vision, at the end of the 1980s, the participatory approach became a key feature in the applications of Communication for Sustainable Development and Social Change (Servaes 2008, 2013, 2014). CSSC is about dialogue, participation, and the sharing of knowledge and information. It takes into account the needs and capacities of all concerned through the integrated and participatory use of communication processes, media, and channels. It works by:

(1) Facilitating participation: giving a voice to different stakeholders to engage in the decision-making process.
(2) Making information understandable and meaningful. It includes explaining and conveying information for the purpose of training, exchange of experience, and sharing of know-how and technology.
(3) Fostering policy acceptance: enacting and promoting policies that increase people's access to services and resources.

The reaction against modernization (and to some extent the realization of global structural imbalances) gave birth to various participatory approaches. They shared the common intent of actively involving people who were the "subjects" of development in shaping the process. But in most cases this is where similarity ends and a diversity of differences begin. People's participation became defined in many different ways and this in turn led to numerous unresolved disagreements (for an overview, see Carpentier 2011; Servaes et al. 1996; Servaes 1999).

Uphoff (1985), for instance, identified four different ways of participation which can be observed in most development projects claiming to be participatory in nature:

(1) *Participation in implementation*: People are actively encouraged and mobilized to take part in the actualization of projects. They are given certain responsibilities and set certain tasks or required to contribute specified resources;
(2) *Participation in evaluation*: Upon completion of a project, people are invited to critique the success or failure of it;
(3) *Participation in benefit*: People take part in enjoying the fruits of a project, this maybe water from a handpump, medical care by a "bare-foot doctor," a truck to transport produce to market, or village meetings in the new community hall; and
(4) *Participation in decision-making*: People initiate, discuss, conceptualize, and plan activities they will all do as a community. Some of this may be related to more common development areas such as building schools or applying for land tenure. Others may be more political, such as removing corrupt officials, supporting parliamentary candidates, or resisting pressures from the elites. Yet others may be cultural or religious in nature–organizing a traditional feast, prayers for an end to the drought, a big party just to have a good time, or deciding on how to implement the SDGs.

Pretty et al. (1995) offers an even more detailed typology of participation:

(1) *Passive Participation*: People participate by being told what is going to happen or has already happened. It is a unilateral announcement by an administration or project management without any listening to people's responses. The information being shared belongs only to external professionals.
(2) *Participation in Information giving*: People participate by answering questions posed by extractive researchers using questionnaire surveys or similar approaches. People do not have the opportunity to influence proceedings, as the findings of the research or decisions are neither shared nor checked for accuracy.
(3) *Participation by consultation*: People participate by being consulted, and external agents or policy makers listen to views. These external agents define both problems and solutions and may modify these in the light of people's responses. Such a consultative process does not concede any share in decision-making, and professionals are under no obligation to take on board people's views.
(4) *Participation for material incentive*: People participate by providing resources, e.g., labor, in return for food, cash, or other material incentives. Much on-farm research falls in this category, as farmers provide the fields but are not involved in the experimentation or process of learning. It is very common to see this called participation, yet people have no stake in prolonging activities when the incentives end.
(5) *Functional Participation*: People participate by forming groups to meet predetermined objectives related to the project, which can involve the development or promotion of externally initiated social organization. Such involvement does not tend to be at early stages or project cycles of planning, but rather after major decisions have been made. These institutions tend to be dependent on external initiators and facilitators, but may become self-dependent.
(6) *Interactive Participation*: People participate in joint analysis, which leads to action plans and the formation of new local institutions or the strengthening of existing ones. It tends to involve interdisciplinary methodologies that seek multiple objectives and make use of systematic and structured learning and participatory processes. These groups take control over local decisions, and so people have a stake in maintaining structures or practices.
(7) *Self-Mobilization*: People participate by taking initiatives independent of external institutions to change systems. Such self-initiated mobilization and collective action may or may not challenge existing inequitable distributions of wealth and power.

Most will agree that participation in decision-making in Uphoff's categorization, or self-mobilization in Petry's typology, is the most important form to promote. It gives people control of their lives and environment. At the same time the people acquire problem-solving skills and acquire full ownership of projects—two important elements that will contribute toward securing the sustained development and the successful and sustained implementation of the SDGs in their community.

Within this framework, communication is viewed as a social process that is not just confined to the media or to messages. CSSC methods (see, e.g., Byrne et al. 2011; Espinosa and Walker 2011; Herman 2016; Hilty and Aebischer 2015; Lennie and Tacchi 2013; Sparks and Rutkowski 2016; Whaleyet al. 2010) are appropriate in dealing with the complex issues of Sustainable Development in order to:

- Improve access to knowledge and information to all sectors of society and especially to vulnerable and marginalized groups;
- Foster effective management and coordination of development initiatives through bottom-up planning;
- Address equity issues through networking and social platforms influencing policy-making;
- Encourage changes in behavior and lifestyles, promoting sustainable consumption patterns through sensitization, and education of large audiences;
- Promote the sustainable use of natural resources considering multiple interests and perspectives, and supporting collaborative management through consultation and negotiation;
- Increase awareness and community mobilization related to social and environmental issues;
- Ensure economic and employment opportunities through timely and adequate information;
- Solve multiple conflicts ensuring dialogue among different components in a society.

2 The Road Ahead

CSSC has been facing new challenges in the last decade, as a consequence of globalization, media liberalization, mediatization, rapid economic and social change, and the emergence of new social media and information and communication technologies (ICTs).

*2.1 * Right to Communicate*

Liberalization has led not only to greater media freedom, but also to the emergence of an increasingly consumer-led and urban-centered communication infrastructure, which is less and less interested in the concerns of the poor and rural people. Women and other vulnerable groups continue to experience marginalization and lack of access to communication resources of all kinds. The issue of ensuring access to information and the right to communicate as a precondition for empowering marginalized groups has been addressed by several meetings and international conferences (World Summit on the Information Society and the World Social Forum).

Today, CSSC programs address global environmental issues, such as biosafety and risk management in disaster prevention and mitigation. Within this framework, communication, education, participation, and public awareness approaches are used in an integrated manner to reach out effectively to the key groups who are needed to protect the environment.

2.2 * *Conserving Environmental Sustainability*

Another challenge faced by rural development programs is how to alleviate poverty and stimulate economic growth while, at the same time, preserving the environment. Fighting land degradation and desertification, halting deforestation, promoting proper management of water resources, and protecting biodiversity require the active participation of rural communities through communication processes.

CSSC focusing on participatory approaches can facilitate dialogue, increase the community knowledge base (both indigenous and modern), promote agricultural practices that are compatible with the environment, and develop awareness among policy makers, authorities and service providers. Furthermore, participatory communication approaches can bring together different stakeholders and enable the poorest and most marginalized to have a voice in the use of natural resources.

2.3 * *Promoting Food Security, Rural Development, and Sustainable Livelihood*

Food security and rural development policies have been revised in recent years placing more emphasis on holistic approaches to rural livelihoods and the sustainable use of natural resources. Furthermore, the spread of digital communication technology has made information and communication services increasingly cost-effective options for providing basic information to dispersed rural producers, in particular to those settled in remote and poorly accessible areas. The focus remains on the needs of rural people, rather than on communication media per se. The critical aim is to enhance the capacity of local stakeholders to manage communication processes, to develop local contents, and to use appropriate media tools. CSSC strategies must be context specific and reflect the values, perceptions, and characteristics of the people and institutions involved.

2.4 * Empowerment of Women, Girls … and Senior Citizens

Communication can also play a decisive role in promoting the empowerment of women and girls with a more equitable framework of gender dynamics. Communication processes allow rural women a voice to advocate changes in policies, attitudes, and social behavior or customs. Through CSSC, women can take control of their lives and participate as equals with men in promoting food security and rural development. In a similar way, the power and expertise of senior citizens, who in many societies enjoy great authority and respect but are often left outside development efforts, should be recognized. Many societies may have to reconsider their contributions and potential anyway, given other social and demographic developments.

2.5 * Narrowing the Digital Divide

The issue of equal access to knowledge and information is becoming one of the key aspects of sustainable development. Vulnerable groups in the rural areas of developing countries are on the wrong side of the digital divide and risk further marginalization. In the rush to "wire" developing countries, little attention has been paid to the design of ICT programs for the poor. The trend ignores many lessons learned over the years by Communication for Development approaches which emphasize communication processes and outcomes over the application of media and technologies. There needs to be a focus on the needs of communities and the benefits of the new technologies rather than the quantity of technologies available (UNDP 2012). Local content and languages are critical to enable the poor to have access to the benefits of the information revolution. The creation of local content requires building on existing and trusted traditional communication systems and methods for collecting and sharing information.

2.6 * Poverty Reduction

Communication can contribute to the effective reduction of poverty and offer better opportunities for the inclusion of marginalized groups and isolated population in the policy-development and decision-making process. Eradicating poverty was the first MDG and remains a priority in the SDGs as well!. Although poverty cannot be divorced from uneven power structures, and communication cannot substitute for structural change, the appropriation of communication for development processes and technologies by marginalized and vulnerable groups, including indigenous people, can ensure that they have a voice in decisions that affect their lives.

2.7 * *Good and Affordable Health for All*

Health should be viewed from a social justice and rights-based perspective. Issues of power and gender, socioeconomic determinants of the problem and both collective and individually oriented responses are equally important to tackle health problems and promote good and affordable health.

The role of the public health analyst or researcher is to assist, not simply a decision-maker, but a decision-making process that has the assent of the community as a whole. In this process the community, as well as the decision-maker and the analysts, are involved.

Only if these key issues are taken into account and addressed in a participatory, democratic, and accountable way, there might be some hope for a sustainable future. However, as argued by the Thai social philosopher Sulak Sivaraksa in the above opening quote, and supported by Chris Goodall, the Winner of the Clarion Award for nonfiction, "we have a moral duty to ensure that future generations—and the most vulnerable billions of our own generation—are not disadvantaged by the pursuit of easy prosperity by people in today's rich world. People put forward many different reasons for taking sustainability issues seriously but it is the moral one that is crucial" (Goodall 2012: 119).

References

Byrne, E., Nicholson, B., & Salem, F. (2011). Information communication technologies and the millennium development goals. *Information Technology for Development, 17*(1), 1–3.

Carpentier, N. (2011). *Media and participation. A site of ideological-democratic struggle*. Bristol: Intellect.

Espinosa, A., & Walker, J. (2011). *A complexity approach to sustainability. Theory and application*. London: Imperial College Press.

Goodall, C. (2012). *Sustainability. All that matters*. London: Hodder & Stoughton.

Herman, R. D. K. (2016, January). Traditional knowledge in a time of crisis: Climate change, culture and communication. *Sustainability Science 11*(1), 163–176.

Hilty, L., & Aebischer, B. (Eds.). (2015). *ICT innovations for sustainability*. Heidelberg: Springer.

Lennie, J., & Tacchi, J. (2013). *Evaluating communication for development. A framework for social change*. London: Routledge Earthscan.

Pretty, J., et al. (1995). *Participatory learning and action: A trainer's guide*. London: International Institute for Environment and Development.

SDSN. (2015, June 22). *Indicators and a monitoring framework for the sustainable development goals. Launching a data revolution for the SDGs*. In: A report by the Leadership Council of the Sustainable Development Solutions Network. Revised working draft.

Servaes, J. (1999). *Communication for development: One world, multiple cultures*. Creskill, NJ: Hampton.

Servaes, J. (Ed.). (2008). *Communication for development and social change*. London: Sage.

Servaes, J. (Ed.). (2013). *Sustainability, participation and culture in communication. Theory and Praxis*. Bristol-Chicago: Intellect-University of Chicago Press.

Servaes, J. (Ed.). (2014). *Technological determinism and social change. Communication in a tech-mad world*. Lanham: Lexington Books.

Servaes, J., Jacobson, T., & White, S. (Eds.). (1996). *Participatory communication for social change*. Sage: New Delhi. 287 pp.
Sivaraksa, S. (2009). *Rediscovering spiritual value. Alternative to consumerism from a Siamese Buddhist perspective*. Bangkok: Sathirakoses-Nagapradipa Foundation.
Sparks, J., & Rutkowski, D. (2016). Exploring project sustainability: Using a multi perspectival, multidimensional approach to frame design. *Development in Practice, 26*(3), 308–320.
UN. (2014). *Millennium development goals report 2014*. New York: United Nations.
UNDP. (2012). *Mobile technologies and empowerment: Enhancing human development through participation and innovation*. Available at: http://undpegov.org
United Nations Secretary-General's High-level Panel on Global Sustainability. (2012). *Resilient people, resilient planet: A future worth choosing*. NY: UN (eISBN: 978-92-1-055304-9).
UNSC. (2015). *Technical report by the Bureau of the United Nations Statistical Commission on the process of the development of an indicator framework for the goals and targets of the post-2015 development agenda (Working draft)*. Retrieved from: https://sustainabledevelopment.un.org/content/documents/6754Technical%20report%20of%20the%20UNSC%20Bureau%20%28final%29.pdf
Uphoff, N. (1985). Fitting projects to people. In M. Cernea (Ed.), *Putting people first: Sociological variables in rural development*. New York: Oxford University Press.
Van Norren, D. E. (2012). The wheel of development: The millennium development goals as a communication and development tool. *Third World Quarterly, 33*(5), 825–836.
Vandemoortele, J. (2011). If not the millennium development goals, then what? *Third World Quarterly, 32*(1), 9–25.
Whaley, D., Weaver, L., & Born, P. (2010). *Approaches to measuring: Community change indicators*. Waterloo ON: Tamarack. www.tamarackcommunity.ca.

Index

J. Servaes (ed.), *Sustainable Development Goals in the Asian Context*, Communication, Culture and Change in Asia 2,
DOI 10.1007/978-981-10-2815-1

Printed by Printforce, the Netherlands